Examining Biophilia and Societal Indifference to Environmental Protection

Mary Ann Markey
Grand Canyon University, USA

Lonny Douglas Meinecke
King University, USA

A volume in the Practice, Progress, and Proficiency in Sustainability (PPPS) Book Series

IGI Global
PUBLISHER of TIMELY KNOWLEDGE

Published in the United States of America by
IGI Global
Engineering Science Reference (an imprint of IGI Global)
701 E. Chocolate Avenue
Hershey PA, USA 17033
Tel: 717-533-8845
Fax: 717-533-8661
E-mail: cust@igi-global.com
Web site: http://www.igi-global.com

Library of Congress Cataloging-in-Publication Data

Names: Markey, Mary Ann, 1949- author.
Title: Examining biophilia and societal indifference to environmental protection / by Mary Ann Markey and Lonny Douglas Meinecke.
Description: Hershey : Engineering Science Reference, [2020] | Includes bibliographical references and index. | Summary: "This book examines biophilia and societal indifference to environmental protection"-- Provided by publisher.
Identifiers: LCCN 2020001938 (print) | LCCN 2020001939 (ebook) | ISBN 9781799844082 (hardcover) | ISBN 9781799853022 (paperback) | ISBN 9781799844099 (ebook)
Subjects: LCSH: Environmental protection--Philosophy. | Nature conservation--Philosophy. | Biodiversity conservation--Philosophy.
Classification: LCC GE140 .M266 2020 (print) | LCC GE140 (ebook) | DDC 363.7001--dc23
LC record available at https://lccn.loc.gov/2020001938
LC ebook record available at https://lccn.loc.gov/2020001939

This book is published in the IGI Global book series Practice, Progress, and Proficiency in Sustainability (PPPS) (ISSN: 2330-3271; eISSN: 2330-328X)

British Cataloguing in Publication Data
A Cataloguing in Publication record for this book is available from the British Library.

All work contributed to this book is new, previously-unpublished material.
The views expressed in this book are those of the authors, but not necessarily of the publisher.

For electronic access to this publication, please contact: eresources@igi-global.com.

Practice, Progress, and Proficiency in Sustainability (PPPS) Book Series

ISSN:2330-3271
EISSN:2330-328X

Editor-in-Chief: Ayman Batisha, International Sustainability Institute, Egypt

MISSION

In a world where traditional business practices are reconsidered and economic activity is performed in a global context, new areas of economic developments are recognized as the key enablers of wealth and income production. This knowledge of information technologies provides infrastructures, systems, and services towards sustainable development.

The **Practices, Progress, and Proficiency in Sustainability (PPPS) Book Series** focuses on the local and global challenges, business opportunities, and societal needs surrounding international collaboration and sustainable development of technology. This series brings together academics, researchers, entrepreneurs, policy makers and government officers aiming to contribute to the progress and proficiency in sustainability.

COVERAGE

- Environmental informatics
- Green Technology
- Socio-Economic
- Global Content and Knowledge Repositories
- Sustainable Development
- Global Business
- Innovation Networks
- Outsourcing
- Intellectual Capital
- Eco-Innovation

IGI Global is currently accepting manuscripts for publication within this series. To submit a proposal for a volume in this series, please contact our Acquisition Editors at Acquisitions@igi-global.com or visit: http://www.igi-global.com/publish/.

The Practice, Progress, and Proficiency in Sustainability (PPPS) Book Series (ISSN 2330-3271) is published by IGI Global, 701 E. Chocolate Avenue, Hershey, PA 17033-1240, USA, www.igi-global.com. This series is composed of titles available for purchase individually; each title is edited to be contextually exclusive from any other title within the series. For pricing and ordering information please visit http://www.igi-global.com/book-series/practice-progress-proficiency-sustainability/73810. Postmaster: Send all address changes to above address.

Titles in this Series

For a list of additional titles in this series, please visit:
http://www.igi-global.com/book-series/practice-progress-proficiency-sustainability/73810

Digital Solutions and the Case for Africa's Sustainable Development
Albert Ong'uti Maake (UNIVERSITY OF LAY ADVENTIST OF KIGALI, Rwanda) Benard Magara Maake (Kisii University, Kenya) and Fredrick Mzee Awuor (Kisii University, Kenya)
Information Science Reference • © 2020 • 300pp • H/C (ISBN: 9781799829676) • US $195.00

Developing Sustainable Food Systems, Policies, and Securities
Abiodun Elijah Obayelu (Federal University of Agriculture, Abeokuta, Nigeria) and Oluwakemi Adeola Obayelu (Unversity of Ibadan, Nigeria)
Engineering Science Reference • © 2020 • 300pp • H/C (ISBN: 9781799825999) • US $195.00

Building an Entrepreneurial and Sustainable Society
Brizeida R. Hernández-Sánchez (University of Salamanca, Spain) José C. Sánchez-García (University of Salamanca, Spain) and Antonio Carrizo Moreira (University of Aveiro, Portugal)
Business Science Reference • © 2020 • 369pp • H/C (ISBN: 9781799827047) • US $235.00

Toward Sustainability Through Digital Technologies and Practices in the Eurasian Region
Gainiya Tazhina (University of International Business, Kazakhstan) and Judith Parker (Teachers College, Columbia University, USA)
Information Science Reference • © 2020 • 294pp • H/C (ISBN: 9781799825517) • US $195.00

Examining the Environmental Impacts of Materials and Buildings
Blaine Erickson Brownell (University of Minnesota, USA)
Engineering Science Reference • © 2020 • 358pp • H/C (ISBN: 9781799824268) • US $245.00

Cross-Border Cooperation (CBC) Strategies for Sustainable Development
Rui Alexandre Castanho (University of Dą browa Górnicza, Portugal)
Engineering Science Reference • © 2020 • 348pp • H/C (ISBN: 9781799825135) • US $225.00

For an entire list of titles in this series, please visit:
http://www.igi-global.com/book-series/practice-progress-proficiency-sustainability/73810

IGI Global
PUBLISHER of TIMELY KNOWLEDGE

701 East Chocolate Avenue, Hershey, PA 17033, USA
Tel: 717-533-8845 x100 • Fax: 717-533-8661
E-Mail: cust@igi-global.com • www.igi-global.com

Table of Contents

Section 2
The Paradox of Environmental Protection

The theme of this section is to inform the reader of the potential that love of nature affords humanity along with the consequences of failing to love nature back.

Section 3
The Cure for Societal Indifference

The theme of this section is to propose a solution to the current environmental crisis.

Foreword

Human civilization has forgotten about the relevance of Nature. Coming from the oldest civilization, the Indus Valley civilization, it behooves me to stress the attitude that Vedic culture held toward Nature. It was an attitude of pure reverence. Vedas referred to this planet inhabited by fauna (*Homo sapiens* being the highest in order), as "Mother Earth." Over time, this attitude has evolved to the extent that modern humans take it for granted as their property, to exploit, desecrate, degrade, at will—regardless of the consequences of such pillage—amounting to sheer molestation of Mother Earth! Ignoring the equal rights of all the other species, humans continue to destroy their natural habitats. From an intelligent species that grew plants for food, humans now raise fellow animals for consumption and that too, without compunction or even remorse! Humans have become blind to the benefits accruing from Nature. Humans have lost the spirit of kinship with their co-inhabitant fauna. In short, the most intelligent species is behaving in an utterly selfish and overwhelmingly greedy manner, destroying Nature wherever it stands, in their unethical ways.

Humans have lost their connection with Nature. The pollution, environmental degradation, disease, droughts, floods, tsunamis, and typhoons the planet is experiencing today are but a consequence of climate change brought about by this feudal attitude! They no longer remember the meaning of living in harmony with Nature. The time has come to remind them of their lack of foresight *vis a vis* Nature. A book on Biophilia is the need of the hour—to stop Humanity from continuing on this destructive path, to retrace their steps back to living in harmony with Nature, to restore the habitats of other species. In short, this species needs to relearn to exist as a part of Nature and not as its owners.

Only a change in attitude can restore the relationship between Man and Nature. This book on Biophilia deals with all aspects that humans have lost track of in their inexorable pursuits and will show them their foibles. It will revive their feelings for Nature, and reeducate them to live better so as to derive mutual benefit from a harmonious relationship with flora and fauna. It will augur well for their wellness and stave off many self-created problems such as pestilence, droughts, war mongering. Given the dedication and abundant concern of Dr. Markey and Dr. Meinecke for planet Earth, I am confident that perusing through all the chapters will correct aberrations in the vision of its readers, accrued over decades, and restore the harmonious relationship between Man, beast, and Nature.

Manjeet Kaur Sharma
Independent Researcher, India

Preface

OVERVIEW

Biophilia means, literally, falling in love with nature. There are many phobias (aversions) and philias (reverences) defined in the field of mental health, to try and explain why some things are loved and some are not. Nature, sadly, too often fits into the category of phobias. There is a need to revive love for Nature primarily because otherwise Nature will perish. This book is about that—why don't humans love the planet anymore? Why has that first love died? How can their cohabitation be restored to one of bliss—for both partners? The Earth seems to be dying, yet human civilization does not love her anymore. Perhaps a feasible way to motivate humans to care again, would be to fall in love again.

To quote a famous scene from a famous book, there is an impediment. That impediment is the angry legal representative of Mrs. Economy, to whom the lovelorn human race is currently wed. Though that current spouse has gone mad and is difficult to love, still, the legal arguments on her behalf are sound. Men cannot just run off and elope with Nature while they are legally bound to their Economy. Something must change if Rochester is to wed his Jane, and save the idea of love for Love's sake, rather than save the idea of Law should govern love. This book about biophilia proposes a way out of this dilemma; the idea of "till death do you part" has a loophole. If one of the parties should perish, the two are free to wed. If either Man or his Economy should perish before the whole of Nature does, Nature is free to remarry. Perhaps the new suitor would actually put Her before his own interests.

AUDIENCE

This book will be of much interest to both laypersons and experts. Many civilian scientists are trying to get grass roots organizations off the ground because they are tired of waiting for formal procedures and will find this book a good addition to their arsenal. Experts will find this book both surprising and a useful resource for their own work. In addition, many career fields will discover that they too have the capability of collaborating with psychologists dedicated to saving the planet from banal human attitudes, and may be more than glad to share their ideas and potential solutions to the challenges everyone faces as inhabitants of this fragile home to so many diverse minds in search of undiscovered miracles.

RELEVANCE

The benefits of this publication are manifold. Foremost, not enough is being done or even committed to in the struggle to save the planet, while most seem unable to see that lost jobs are a minor worry compared to a dead planet. Second, it should matter because the authors have research showing that opposition to, denial of, and refusal to restore the Earth to its natural state seem mainly due to an odd phenomenon the authors call the *Hurricane Survivor Effect*. This effect is simply the surprising difference between those who have lost everything versus those who have lost a little. Those who have lost everything except each other are deeply grateful for their survival; those who have lost even a little are upset at having to make repairs. Because the human species does not realize it is about to lose everything, it becomes furious when asked to give back a little (unless its neighbors go first). The authors believe the themes in this work will dramatically expand the field by spurring both sides to action—and the resulting publicity will help raise awareness for the urgency of this proposition.

It has become obvious that simply "crunching the numbers" (converting natural resources into sustainable profit and loss statements) does not communicate the critical "tipping point" looming so close at hand. The human species is approaching a point of no return in reversing the deleterious effects that an invasive, over-colonizing species has had on a once unbeaten, untied, unscored upon Earth. However, challenging everyone's thinking and behavior in ways never attempted before, taking a second look at Nature as a companion not a competitor, making her the Bride and not a concubine, could shift the human mindset into the achievement of a common goal: to save the planet and every species within the same biosphere from certain extinction. Biophilia is more than falling in love just once. It is the incapacity not to fall in love every day with the same woman. Her name is Terra. She needs our contrition dearly.

ORGANIZATION OF THE BOOK

This book is organized into three sections of four chapters each, for a total of 12 chapters. Each section addresses one of the three topics suggested by the title of this book. A brief description of each section and chapter follows:

Section 1 proposes the need to know about and fall in love with biophilia again. These first four chapters introduce the reader to biophilia, and examine the importance of this concept to the human species. The basic and expanded theories of biophilia are covered, as are several key prejudicial attitudes that these theories address. The side effect of classifying everything is investigated in depth, because there is growing evidence that knowledge is not very fond of awe. As if scorn of natural species were not enough, the scorn of certain stages of the human's own species is proceeding apace, with childhood scheduled to be deprecated, and old age relegated to a temporal refugia where those who survived middle age struggle to eke out an existence on the frontiers of economic growth.

Chapter 1 begins by defining the primary concept of this book, biophilia. It exposes the reader to a theory about love for Nature on this world, and the interconnectedness of all species. It describes the odd history of the term biophilia, and explores the definition of love. The potential for Nature to cure human diseases is covered, as are topics such as the exclusion of vibrant species from the idea of sentience and compassion by civilized societies, but never by "savage" tribes.

Chapter 2 takes biophilia to the stars, with a theory about love for Nature beyond the world as humans know it. The importance of not carrying human prejudice along when discovering other worlds is covered, as is the reminder to carry awe for Nature into the cosmos. A new definition of Life is offered, and the benefits of bringing along non-human companions from Earth are explained.

Chapter 3 introduces the problem with scientific classification, and how that rigorous goal may underpin an unfortunate loss of regard for each categorized thing. Concepts like good and evil, like their odd medical counterparts ("health" and "disease") may stem from this same habit of aggregating unique things into convenient heaps. Eastern philosophies are contrasted with Western beliefs, and famous paradoxes and ladders of existential worth are contested. The very idea of species is challenged.

Chapter 4 raises an alarm regarding the mistreatment of those who are too young to work and those who are too old to work, and appeals for a return to a deep and abiding reverence for both childhood and old age. Every stage of the human lifespan matters; every moment matters. An appreciation for Time - no matter how brief - is discussed in this chapter.

Section 2 concerns the paradoxes that go along with "environmental protection." These four chapters describe the mutually rewarding benefits of reverence for Nature that operate in both directions. They also cover the consequences of the current habit of revering people and their cities only. On one hand, there are very good reasons to preserve the environment, because simple exposure to Nature is like a natural Huxleyan *soma* that calms stressed individuals without having to ingest anything to relieve their suffering. On the same hand, a transformation of the planet back to its primal state might be the very thing the climate and its animals (including this human animal) need to survive. On the other hand, recent offers to protect the environment look a lot more like neocolonialism than compassion, offering mainly to raise funds rather than to raise awareness or action. Perhaps humanity has reached a tipping point, and its attitudes are about to go into a spin from which it will not recover. The belief that more sovereignty is the solution when less sovereignty fails is brought into question. Perhaps this is that long-awaited juncture in history when humility and faith, not arrogance and land management, will win the day.

Chapter 5 discusses the healing potential of biophilia. It describes the many benefits of time together with Nature instead of time with consumer products and rampant commercialism. Forest therapy is covered, and a therapy for loneliness is offered, wherein instead of withdrawing from Life, one finds respite in it. Green spaces are discussed, and the distorted perceptions and habits of humans toward food as a right of their species as though there were no consequences for Nature. Natural labyrinths are mentioned as well, as both attractive and comforting, in addition to the concept of seeking out sacred spaces.

Chapter 6 proposes the first of several dramatic solutions to the current crises—*National Park Theory*. If humans have misplaced their heart, perhaps they can find it out in Nature? The science behind lasting mental impressions is explored in depth, and the neuroscience of regard is ablated for the reader to ponder the ramifications. Feedforward thinking is explained, divine hiddenness is explored, and the need to matter is laid bare. Lastly, the eerie equivalence between how humans disregard both their physical body and their physical world—while worshipping their mind and its mental territory—is made plain.

Chapter 7 delves headlong into the murky depths of environmental protectionism, wherein the mute and vulnerable are aptly exploited to raise capital for lobbyists and lawmakers who dwell in forests of marble and alabaster, while their impoverished constituents dwell out in the shivering or beheaded forests they represent. Protection based on conditions of submission and tribute are contrasted with unconditional love (love no matter what). Disbelief in altruism is covered, as is the odd perception that the concept of beneficence is more about ethics than it is about grace. An actual case is explored, showing how industries appeal to the public using national wonders as bait, but with the primary intention of saving their industries, not natural wonders. Fruition, guardianship, and colonialism are explored at length.

Chapter 8 offers the reader a choice—increase our sovereignty over living things, or yield sovereignty back to its rightful owners. The unpleasant truth about historical and modern wildlife management is discussed as is the tragedy of the human inability to comprehend natural population and habitat control. The outcomes of human interventions into (and hyperrational approaches applied to) anthropogenic problems are laid out in the form of habitat destruction and global pollution—the latter of which had no natural precedent till humans came along. Many examples of toxic outcomes are explained in full (toxic to Life but not to industry). The biophilic theory is applied and optimism restored before the chapter and section close—just in time to propose a cure.

Section 3 is all about the cure. Mankind is indifferent to the death of all Life as he knows it, primarily because he thinks that if he simply does not believe in it, it won't happen. This section challenges the reader to reach deep—not into his or her pocketbook for donations that plants and animals cannot spend, but deep into his or her psyche to pluck out the thoughts that keep Nature out of paradise and wailing in hell. The last section does not disappoint; no call to action is much use without a direction in which to direct that action. For this reason, the authors offer not one but several proposals regarding the possibility of averting planetary extinction. The psychology of losing everything is explored; the psychology of counseling irreconcilable partners is investigated. And the authors bring to bear some of the most powerful parables ever shared whose time seems to have come. If the greatest two-legged animals cannot save the world, perhaps the smallest one-stem flowers can? Perhaps the one thing humans have not thought of, is believing what is already here, rather than trying to create things to believe in. No more into the breach dear friends; instead this mental wall between Man and Nature must be pulled down, this mental veil rent in two—so that both sides of the same holy city may enjoy the right to the same freedoms and the same Divine regard.

Chapter 9 proposes the first of several cures for the environmental crisis and societal indifference to that crisis. This first cure is a theory, but it is a theory with quite a bit of evidence. When a man stands to lose a little, he gets upset; but let a man lose everything (except the people he loves), and he thanks God for his good fortune. This is called the *Hurricane Survivor Effect* by the authors. Extreme gains appear to require extreme losses before they become possible. This phenomenon may explain the altruistic behaviors that others think impossible. This effect may explain the *schadenfreude* response (joy at another's sorrow). Pair-bonding is covered; trans-species psychology is explained, and the evidential devolution of a wild mercy into organized *theriocide* is treated. DNA and a common ancestor are discussed, as is the possibility of an affection that often crosses the genetic frontier to bond creatures of different species as though they shared one cherished lineage.

Chapter 10 is all about conflict resolution, which is the expertise of the primary author. Environmental disputes are explained, as are family mediation, Nature's rights, and the proposal of a new, biophilic mode of conflict resolution.

Chapter 11 proposes the second cure for a dying planet. How can reverence for a flower save the entire world? The psychology of reverence for things that do not expect reverence versus those that not only expect but demand reverence is covered. The idea of what is precious, what matters most, and what is unlike anything else in the entire world is analyzed for its potential to incentivize unusual risk-taking—perhaps sufficient courage to save the world. Gratitude for the transitory versus scorn of ephemera (as they apply to the heart of Man) heart are covered in depth. The earlier chapters on childhood and fruition are connected to this key psychological incentive and their power over hubris. The Nature of memory is explored and how memory may discourage saving the very thing human memories seem to preserve. And last, advice from thousands of years ago seems to have found its niche, in that the wisdom lost on those deaf ears may at last find good soil in modern eyes.

Chapter 12 proposes the third and most significant of the potential planet-saving treatments. Could it really be this simple? Is the only thing standing between the death of the planet and a miraculous cure the overdue humility of just one stubborn species? Just as the *Theory of Astrophilia* took the reader beyond the bouys of biophilia, the idea of no greater love asks the reader to venture out beyond the boundaries of first love—to lay down these rights and possessions once and for all, for everyone's Beloved. With both hands *free at last*, men may at last reach out with all they have and grab their green-eyed Bride before she collapses; and though this vain and uncertain species perish in that effort, still it may carry Her to that bright and certain future it so gladly reserved for Her—to safety, to faith, to Life.

Acknowledgment

Dr. Markey and Dr. Meinecke would like to gratefully acknowledge all of the assistance and support that was provided to them in the form of comments, suggestions, feedback, proofreading and technical support, as well as in the belief that the publication of this book would make a significant contribution to the perspective being advanced by biophilists and astrophilists around the globe.

These authors would also like to acknowledge all of the love, patience, and understanding that was demonstrated by their family members along the path to this book about biophilia: Mr. George Markey (husband of Dr. Mary Ann Markey), Mrs. Marcia Perez Meinecke (late wife of Dr. Lonny Meinecke), Mrs. Lisa Moore Meinecke (wife of Dr. Lonny Meinecke), and Jenny Nicole Meinecke (daughter of Dr. Lonny Meinecke).

Introduction

This book is about three things: biophilia, societal indifference, and environmental protection. Biophilia is simply the profound love a living thing has for physical Life, even when its lifespan is brief. Societal indifference is the ambivalence for Life peculiar to this mentally reverent species called *Homo sapiens*, even when its lifespan is long. Environmental protection is the newest in a long line of cognitive façades, beneath which façade hides the true nature of cognition (theft and derision of everything it isn't), as evidenced by an unwavering struggle for global sovereignty. The authors submit this view because civilized people typically regard the less civilized as inferior to their civilization, and offer aid only if some form of subjugation is involved.

The authors have set out in this volume a theory. The crux of this theory is the sublimation of sovereignty with faith. In psychology, sublimation is the transformation of undesirable, pathological impulses into desirable, prosocial impulses. The authors (both of whom hold doctorates in psychology), hypothesize that if the demand for sovereignty over Life of every kind throughout recorded history has failed to preserve so much as ruin lives long before their time, perhaps an investment in reverence for Life of every kind will succeed in preserving the few living species that remain? Such a faith in scientific terms is simply to trust in the non-manipulation of outcomes. Most call it Chance (because they are afraid of Nature and of one another). A few operationalize it as Faith in a Nature which happens to include one another.

A brief explanation of how this transformation of reluctant human attitudes might come about would probably read something like, "Do you love me? Then *show me* how much you love me." It has been said by authors more able than these, that there is no greater love than a love that puts its Beloved before itself. That is the cure for indifference to the death of the planet these humble scribes offer their field; this is the solution to saving a dying Bride.

Section 1
The Need for Biophilia

This introduction to the first book section defines the purpose of this first part. This part is entitled "The Need for Biophilia." The four chapters included in this section include "The Theory of Biophilia," "Beyond Biophilia: The Theory of Astrophilia," "How Taxonomy Steals Reverence," and "Reverence for Childhood and Old Age." This introduction simply outlines the themes that the first four chapters will cover, serves as one of the dividers of the material into three logical chunks, and helps outline how they fit into the organization of the book and its three objectives.

The purpose of this book is to help humans, as a natural species, belatedly admit to ignorance when it comes to the value of other natural species, so that by a similar misfortune of Nature, this species may find its reverence for Life. If philosophy begins with the admission of ignorance, then psychology is surely the analysis of that ignorance (Plato, & Stock, 2010). This book, then, will set out to not only admit to a neglect of its vulnerable companion (this home of homes called Earth), but introspect what it has been neglecting, why it has been neglecting it, and how it may go about transforming its antibiotic impiety and bias to one representative of a more probiotic humility and mercy.

The theme of this first section is to provide a background of the problem. The planet is dying. The human species is poisoning it. Humans are the thief of Earth's joy, to borrow from a phrase attributed to Theodore Roosevelt (Cooper, Gustafson, & Salah, 2013). Notwithstanding, no hypothesized problem is complete without a theorized solution, so this first section will also outline the antidote: reverence for Life in all its kinds. There is a word for it already . . . humans call that word, *biophilia*.

REFERENCES

Cooper, K. B., Gustafson, N., & Salah, J. G. (2013). *Becoming a great school: Harnessing the powers of quality management and collaborative leadership*. Lanham, MD: Rowman & Littlefield Education.

Plato, & Stock, S. G. (2010). *The apology of Plato*. Charleston, SC: BiblioLife.

Chapter 1
The Theory of Biophilia

ABSTRACT

This first chapter introduces the Theory of Biophilia, which argues that a crucial relationship exists among all species, both human and non-human. This theory is unique in that it emphasizes an almost reverent collaboration of all life on the planet Earth toward a common need to survive, rather than their apparent hostility and competitiveness. The Theory of Biophilia expands the human definition of sentient life and examines how important this love of life is to the essential need for regard within both the observer and the observed.

INTRODUCTION

What is *biophilia*? A little background on a big deal like biophilia seems like a good place to start this book. The first occurrence of the term still used today surfaces around 1857, but strangely lumped together with mental disorders like hypochondria, melancholia, and the like (Jahr, 1857). What? Apparently, biophilia was first likened to an illness in which the patient seemed too mindful of his or her body states, rather than being more mindful of his or her mental states. That does sound like an illness—but moreso for the definition than for the disease. Biophilia suggests fondness for biological things (like the body), whereas fondness for Life and living things is hard to do if you are busy loving stuff in your head instead. The original source goes on to state that the chronic condition (biophilia) was thought to result in "unwanted" thoughts about primarily somatic worries, such as the possibility of physical sickness or death (Jahr, 1857). Really? Curiously, this scorn for annoying

DOI: 10.4018/978-1-7998-4408-2.ch001

physical issues back in the past was not so very different from "unwanted" thoughts about the impending death of the planet, which many write off as an annoying group delusion.

By 1892, biophilia had become a more specific disorder in a manual of medical and mental disorders, which manual defined biophilia this way: "The instinct of self-preservation common to man and the *lower animals*" (italics added; Tuke, 1892, p. 135). Well, this is an improvement—now persons hampered by the condition known as biophilia were at least as intelligent as the lower animals, if they were not otherwise sick. This does align, however, with an ongoing view of Nature as *endearing yet inferior*, especially its animals. Interestingly, this same view (which by the by, was held by the same gender that wrote all the books) was also held toward women and children—and just about anything that wasn't old and gray and of the male persuasion (Hodson, Kteily, & Hoffarth, 2014; Meinecke, 2017). After 35 years, concern for living things had only improved from an inconvenient human hypochondria to a vestigial animal instinct.

It was Fromm (1964) who must have drawn from these earlier works (the wording is very similar), by elevating the erstwhile vestigial drive to a well-known Freudian concept known as the *life (or love) instinct* (Gerber, 2019; Maizels, in press). Again—curiously—comparing the love of living things to the life instinct and global war to the *death (or hate) instinct*[1], was very like the current environmental ethics debate, in which some wish only to protect life (the love of life instinct), and others to simply profit from it (a sort of hateful jealousy of life instinct; Nordstrom, 2004). One can readily see the term (biophilia) lifted up from its prior disease model, much like psychology itself has gone from an illness model to a wellness model (even if the focus is still on profit). By 1964, biophilia had begun to refer to a genetically natural impulse overshadowed by recent (and more dominant) drives—which, at times, surfaced as a recessive gene more similar to a pathological impulse than a forgotten empathic gene (Fromm, 1964).

By 2006, biophilia was neither a disease nor a positive attribute of the human species. Like synesthesia, biophilia had become a juvenile trait or "green gene" that survived childhood in women but not in men (Ramachandran & Hubbard, 2001; Weir, 2006). This might be teleologically explainable by the onset of spermarchy and the fierce competition for mates, if viewed through an *evo-devo*[2] lens. Then, around age 45, secure in the survival of his heredity, the male might let slip the dogs of pacificism in his green genes once again, given the arrival of middle age and the inexorable waning of his previously hot and waxing libido (Buss & Barnes, 1986). Meanwhile, under the radar so to speak, in 1979 the famous biologist E. O. Wilson had proposed that biophilia might be a hidden need that incidentally flourished in the presence of living things. By 1984, Wilson had composed a booklet under the term, suggesting that biophilia was an "innate human tendency to focus on and affiliate with life forms and life-like processes" (Joye & De Block, 2011, p. 190). This idea was met with (and is still met with) much resistance, perhaps in proportion to its continuing popularity, as most popular successes seem to evince. Even 10 years later, Gould (1994) was still arguing that the human propensity for creatively destroying itself was equal to its propensity to creatively grieve over what it was destroying.

Today, biophilia has become more a question of which ethics one finds continuing commercial benefit from, rather than from an (unprofitable) reverence for Life (Joye & De Block, 2011; Meyer & Bergel, 2012). At present, biophilia has become a public sensation and an industrial boon—even if Nature herself is no closer to being rescued from human expansionism than Wilson's futile hope that humanity will (willingly) give back half the planet to create an unsustainable partnership with its minority shareholder and silent partner[3] (Wilson, 2016). The concept has spawned numerous publications, new fields within science, medicine, academics, and psychology, and even spawned an unexpected reception among eager copycats and bureaucrats of Nature's surviving species, with products like "biophilic architecture" and careers in "environmental science" (Joye & De Block, 2011). In the true fashion of opportunistic-thinking, biophilia is no longer a chronic vigilance to the body's suffering under the lashes of modern pyramid builders, nor the horrified admission of a conscienceless genocide of the former inhabitants of the planet Earth. No, biophilia is now a cognitive commodity with tremendous economic growth potential, along with a bevy of self-gratifying theories (well-received by a mutual admiration society), suggesting only species hominin cares about saving the environment from species hominin.

Once upon a time, though, biophilia referred to an endless fondness for living things, much as the deepest love for anybody begins as fondness. That's because the Greek words that form the modern term *bio-philia* once stood for life (*βίος*) and fondness (*φιλία*), according to scholars and a wealth of surviving ancient writing (Perseus Digital Library, 2020). The problem with *fondness*, however, is twofold (and *love* has always been a problem to define). The first problem is that there are so very many things one might become fond of, but one can only be fond of one thing at a time.

The second problem is that there are two kinds of "things": (a) naturally occurring things that were already here (i.e. not invented by Man), and (b) unnatural things made from the unhappy remains of natural things (i.e. "invented" by Man). The second problem is really just a question of which *kind* of thing you love more—a living thing (when you are driven by the love of life instinct), or a non-living thing (when you are driven by the love of death instinct; Nordstrom, 2004). Hopefully the prefix "bio" prepended to "philia" helps solve the question of which kind of love this book on biophilia is attempting to describe. Incidentally, blood sacrifices suggest the love of death, since previously *living* things are offered up to placate Death by those who fear Death (Gilhus, 2006).

The first problem is not so easy to solve. What should humans love first? Better said, since things can take turns at being first, what should humans love most of all? In that one dilemma lies the *hardest problem of consciousness* (Chalmers, 1995). The best Greek philosophers could not solve it. So they divided the complete idea of love (lifelong fondness) into several incomplete ideas, so that each incomplete idea of love could have its own first love—much as any person today struggles to view his or her personal hormone zoo of perfect partners as his or her "one true love." According to Birch (2019), a possible list of seven kinds of love in Greek thought and practice would include:

1. Eros: romance and passionate love
2. Philia: brotherly love or an inseverable friendship
3. Ludus: mischievous or flirtatious love
4. Storge: parental love
5. Philautia: self-love
6. Pragma: practical (logical, sensible, utilitarian) love
7. Agápe: selfless love (the highest form of love or charity)

But no matter how many times humans divide love, if humans do not love the same thing day in, and day out, whatever humans used to love will become either melancholic or mad when the shadow of love no longer falls on the beloved object at the proper time (Freud, 1917). As to the melancholic outcome, if the previously loved object had low self-esteem or low SDO[4], it will become overwhelmed with sorrow and miss being loved yesterday (Marchant, 1646; Spinoza & Curley, 1994). As to the mad outcome, if the previously loved object had high self-esteem or high SDO, its spurned expectations will swell with a jealous rage and demand being loved tomorrow (Jylhä, Cantal, Akrami, & Milfont, 2016; Meinecke, 2018a, 2018b; Skinner, 1957).

This is the case with the human species and its love affair with progress—humans are either melancholic or mad about the environmental crisis. Those who are melancholic exhibit the love of life instinct (*das Es* – their prior need for pleasure as an affirmation of being). Those who are mad exhibit the love of Self instinct (*das Ich* or *das Uber Ich* – their new need for dominion as an affirmation of being). Humans have forgotten their first love, and run off with a strange painted woman whose flesh will never wrinkle and whose hair will never turn gray (Prov. 5: 1-5). The love of flesh and blood that humans once thought sacred (biophilia), is now a savage inclination, or a social taboo (Freud & Brill, 1913).

THE THEORY

In a nutshell, the *Theory of Biophilia* addresses the innate relationship that exists among all of the species that share the biosphere on Earth; however, what makes this theory unique, is that it assumes a *collaborative mindset* (a partnership for survival), rather than an *adversarial mindset* (a competition for world domination). The Theory of Biophilia is posited as follows:

Within the earth's biosphere, all species both non-human and human are inextricably linked and innately motivated to consistently focus on and to interact with other species. When these motivations to focus on and interact with other species demonstrate the characteristics of awe, reverence, respect, and/or empathy, these inter-species interests and interactions can be mutually beneficial psychologically, biologically, emotionally, and spiritually in ways that encourage the species' survival, evolution, development, and ability to flourish. Conversely, when these inter-species interests and interactions exhibit characteristics that are indicative of egocentrism, self-serving biases, devaluation, and domination of one species by or over another species, the outcomes can be catastrophic not only for that specific interspecies exchange but also across the entire biosphere resulting in the decimation, destruction, and/or extinction of both known and yet-to-be discovered species. (O'Grady, 2016)

So, where to begin when there is no indication of where to start, or even what questions to ask? Even the lyrics from the song "Do-Re-Mi" from *The Sound of Music* (Wise, 1965) advise starting at the beginning—because the very beginning is "A very good place to start" (Rodgers, 1959, stanza 2).

Even the examples given are from Nature, such as: "Doe, a deer, a female deer" (Rodgers, 1959, stanza 10) and "Ray, a drop of golden sun" (Rodgers, 1959, stanza 11). These examples suggest that contemporary music itself reflects an intimate connection with Nature (Suskin, 2009).

Make no mistake that this intimate, innate, interspecies connection is anchored very securely across multiple contexts . . . including biology, psychology, sociology and spirituality . . . which is evident when you assume a biophilic perspective. It is this shift in human perspective when it comes to interspecies relationships (from biophilic to *biophobic*), that has changed drastically over time. This is particularly so as it relates to the human versus non-human relationship, as the escalation of natural signals portending dire consequences flood the news today. Despite the inevitability of such irremediable consequences, the evidence of human disruption of vital ecological systems continues to be denied or ignored. Included among the challenges plaguing the Earth's biosphere today are:

- Multiple types of pollution (air, water, and light)
- Consistently more extreme weather patterns and storms
- Climate change
- Rising ocean levels
- Melting ice floes
- And the list goes on

DECONSTRUCTING A THEORY: AWED BY NATURE

Scientific investigation generally begins with definitions, terminology, and the operationalization of factors that are an integral part of the research or the initiative. In this case, the term, *biophilia,* according to the Cambridge Dictionary, may be defined as the innate affinity humans have for other forms of life (Biophilia, 2019). The online website at Dictionary.com also provides a similar definition for *biophilia* as a love of life (with no restriction of the definition of love to the human species), which raises the question as to whether other non-human life forms might also demonstrate an affinity for other non-human and human life forms (Biophilia, 2020). In response to that question, biophilia researchers and practitioners reply with a resounding "Yes!" The definition goes on to also mention the affinity that humans have for other life forms. The Theory of Biophilia specifically states that the motivations underlying interspecies interactions should demonstrate the characteristics of awe, reverence, respect and/or empathy, since certain situations or circumstances may elicit a stronger response of one characteristic over another.

Awe may be characterized as an emotion that is associated with feelings of wonder that are inspired by a source that is deemed to be sacred or sublime (Awe, 2019); however, these authors further suggest that "awe" is a feeling that arises when words fail to provide an adequate response (i.e. awe renders people speechless).

Reverence is a term that is often associated with the recognition of one's right to be loved by or devoted to something, while *respect* suggests the act of giving specific attention, holding in high regard, and regarding with special consideration (Respect, 2019; Reverence, 2019).

Empathy, according to *Psychology Today* (2019), is the experience or ability of an individual to understand the thoughts, the feelings, the situation, and the subjective condition of another—rather than from one's own point of view. This ultimately promotes prosocial behaviors. These "helping" behaviors arise from within intrinsically, in contrast to being coerced or forced to be extrinsically compassionate. Research has suggested that, although there are individual differences in the ability to empathize that are genetically based, it is possible to increase an individual's capacity for empathetic understanding.

Justin Bariso (2019) has suggested that there are actually three types of empathy that can aid in building healthier and stronger relationships: cognitive, emotional and compassionate. *Cognitive empathy* enables individuals to comprehend how someone else might be feeling and what they may be thinking while *emotional empathy* or *affective empathy* enables individuals to share the feelings of others. *Compassionate empathy* or *empathetic concern* moves beyond comprehending others and sharing their feelings, because it compels individuals into action in order to help in any way possible.

DECONSTRUCTING A THEORY: ANIMAL SENTIENCE

Current research also indicates that non-humans not only have the capacity to empathize, but their behaviors also demonstrate that they have a moral compass (Morell, 2013). According to National Geographic, Carl Safina who is an environmental journalist documents this phenomenon in his book entitled, *How Animals Think and Feel* (2015), in which he details how the acknowledgement of consciousness in non-humans as well as new discoveries about the functioning of the brain have served to diminish the barriers between humans and their non-human counterparts. It has become imperative that humans learn much more about the thoughts and feelings of other species with whom they share this biosphere, and divert from their usually number-crunching perspective as it relates to conservation efforts. Humans are dismayed when they are regarded as "just a number" in the sea of their human counterparts, and *so it is* with non-humans—who perceive existence in a very vivid fashion (Grandin, 2005). Non-humans also live in the present, where they too focus on identifying their friends and foes, and compete for higher status and resources that will allow them to live and contribute progeny for future generations (Mendl & Paul, 2008). In other words, their presence here on Earth enriches the quality of life for all species with whom they co-exist, and it appears that their human counterparts have been lagging behind in that realization—until a ray of illumination broke through with the recent publication of an article in the *American Association for the Advancement of Science*, which investigated the "culture" of fruit flies. The researchers Danchin et al. (2018) discovered that these insects are able to engage in observational learning from the other fruit flies around them, and then pass those behaviors and preferences on to others. They conducted one experiment in which observer female fruit flies were positioned where they could watch demonstrator female fruit flies choose between two males that had been colored pink or green. Later the observer female fruit flies chose the same color of male fruit fly as a mate that the demonstrator females had chosen more than 70% of the time. In a separate experiment with the progeny of the observer fruit flies, it was noted that these color preferences could be transmitted down to the eighth generation of their offspring.

As evidence for the existence of animal sentience, consciousness (as well as feelings of empathy and love in non-humans) continue to be documented (Allen-Hermanson, 2018). Humans are confronted more dramatically and more often by moral and ethical dilemmas relating to the use of non-humans for the purposes of basic research that benefits humans alone, whereas in stark contrast, using humans for basic research is strictly forbidden (no matter who benefits). A recent letter-to-the-editor by Michael Radkowsky in the American Psychological Association's publication, *Monitor on Psychology,* is directing that query squarely into the lap of the former president of this prestigious organization:

I am writing regarding former APA President, Antonio Fuente's column "The Importance of Research with Nonhuman Animals (October 2017). Given that non-human animals are, like humans, sentient beings who experience fear and pain, how can we justify subjecting them to experiments that certainly cause them fear and pain? Simply because we have the power to do so? Yes, the findings of research done on nonhuman animals may benefit humans, but does that make such research morally right? APA, which should be the standard-bearers of ethics for psychologists, should not justify and support inflicting psychological suffering on other sentient beings. (Radkowsky, 2019)

Authors Murphy-Gill and Johnson (2016) weighed in on the discussion of this topic in an article they published on the U.S. Catholic website, *Do dogs go to heaven?* in which they presented an historical perspective of the roles that pets have assumed in the lives of humans, as well as a discussion of the Pope's position on the subject, which ultimately impacts the Church's position regarding the possibility of non-humans having souls. Centuries past, humans kept animals around for more utilitarian purposes, such as hunting, pest control, and general working tasks; however, according to a 2015-2016 survey that was conducted by the American Pet Products Association or APAA, approximately 79.7 million American households are home to a pet (American Pet Products Association, 2015). Human pets are now typically elevated to the status of being categorized as a companion animal, in contrast to having to earn their keep (service animals). Since humans hold their human and non-human companions in lovingly high regard, it seems logical that humans would want to know what happens to them after they die, and where they fit into God's overall scheme of things.

Over the reign of several Popes since Pius IX, the Papacy has held differing views regarding the condition and treatment of non-humans; however, Pope Paul VI was reported as having consoled a boy whose pet dog had died by saying, "One day we will see our pets in the eternity of Christ." In 1990, Pope John Paul II proclaimed to a papal audience that "animals possess a soul and men must love and feel solidarity with our smaller brethren." He also added that animals are the "fruit of the creative action of the Holy Spirit and merit respect," and he also commented that they are "as near to God as men are." So, although Pope John Paul II never claimed that animals would enter the kingdom of heaven, he enforced the concept that all animals or non-humans are God's creation, and therefore, worthy of love and respect. Pope Francis—named after the patron saint of animals who famously welcome all animals as fellow creatures of God—prayed in his encyclical on the Earth's environment:

Teach us to discover the worth of each thing, to be filled with awe and contemplation, to recognize that we are profoundly united with every creature as we journey towards [God's] infinite light.

So, while it may not be possible for humans to determine if their beloved pets will go to heaven, there is sufficient certainty that non-humans are fellow creatures created by God, and therefore, worthy of love and dignity. Pope Francis remains somewhat of an environmental activist. When contacted by this author for his blessing on the First International Biophilia Conference that she was co-hosting in June, 2018 at the Mount Melleray Abbey in the rural mountains of Waterford, Ireland, the Secretariat of State from the Vatican responded with this personal communication dated 20 August 2018:

Dear Dr. O'Grady

His Holiness Pope Francis has received your letter, and he has asked me to respond. He thanks you for writing to him about your efforts to care for the world, our common home.

The Holy Father will remember your intentions in his prayers, and he invokes upon you God's blessings of joy and peace.

Yours sincerely,
Monsignor Paolo Borgia
Assessor

Changes in the doctrines of the Catholic Church tend to move forward very slowly, so such comments uttered by Pope John Paul II and the current Pope Francis regarding all God's creatures, may have profound implications for the status and treatment of all species. In any case, these amazingly open-minded papal statements echo a new definition for what charity is and what charity is not. In an address to Caritas Internationalis, the meaning of *charity* (charity is the word for love in the King James Version of the Bible), is that of a simple relationship between one another, not of a formal business transaction. The statement added that " . . . charity must involve the heart, the soul and our whole being" (Vatican News, 2019, para. 5-6). Like biophilia, charity is not a *for-profit social enterprise*, because the reward is in its grateful exercise, not in its compensation (Carson & Griffith, 2009; Matt. 6:1-4).

BIOPHILIA: A MEANS TO CURE DISEASES

At first glance, conducting an investigation as to how biophilia functions within a biological context does appear to be rather redundant; however, one of the more unique research labs in the world is run by Dr. Bryan Grieg Fry (2019) who is an expert in the field of venom and toxins. His research extends far beyond the typically medically significant reptiles, scorpions and spiders to the point at which he studies all venomous animals, such as sea anemones, jellyfish, sea snails, cephalopods, centipedes, some orders of insects, echinoderms, fish, and certain mammals, such as lorises, platypus and shrews. Dr. Fry's research program employs a multiple-disciplinary approach that integrates ecological, evolutionary, and functional genomics in order to gain a better understanding of the evolution of venom systems. His major findings currently indicate: (a) venom systems are more widespread than previously thought; (b) convergent strategies are needed for the detailing of the functional and structural constraints on those proteins that are recruited for use as venom toxins; (c) there is a need to redefine what makes an animal venomous; and (d) there seems a revelation of a greater diversity of animals with toxins that can be utilized in drug design and development. He investigates how venoms and toxins can be employed as the lead compounds in drug R&D (research and development), and he has determined that the most divergent venoms and toxins will have the most unique compounds which are most likely to be useful as therapeutics.

One of Dr. Fry's critical life-saving research projects concerns which antivenoms are the best match for snakebites, particularly regarding the geographical region of that venom for its effects upon the blood. His work included the Saw Scale Viper of the genus *Echis*, which is a leading cause of snakebite morbidity and mortality in Sub-Saharan Africa, the Middle East, and certain regions of Asia, and which constitutes a public health issue that exceeds almost any other snake genus on a global level. This study addressed a comprehensive testing of *Echis* venom effects upon the coagulation of human blood plasma as well as the most extensive examination of antivenom potency and cross-reactivity up to this point. *Echis* venom predominately consists of hemotoxins and cytotoxins which usually cause coagulopathy and defibrination that can persist for days even weeks, and which can result in bleeding anywhere in the body—including intracranial hemorrhage—with the latter occurring a few days after the bite. Test results indicated that venoms are generally potently procoagulant, but there is a high variability of the relative

Figure 1. Dr. Fry holding a Komodo dragon
(© 2013, Dr. Bryan Fry. Used with permission)

dependency on calcium or phospholipid co-factors. In addition, three of the four antivenoms exhibited a very narrow taxonomic range of effectiveness to prevent coagulopathy. Dr. Fry's findings inform the field regarding the potential clinical effects of envenomation in humans, as well as highlight the extreme limitations of the treatment available. This is part of an effort to develop antivenoms that have a more comprehensive impact for snakebites from snakes in the wild as well as from species that are widely kept in zoological collections (Rogalski et al., 2017). In addition, research involving venoms and toxins fosters the development of new and improved medicines for the treatment of medical issues, such as hemophilia (Henderson, 2017) and heart attacks which were advanced through the investigation of the Russel's Viper (*Daboia russelii*). Those included Rapaport, Aas, and Owren's (1954) seminal study, and the Wagler's Viper (*Tropidolaemus wagleri*) respectively. According to Dr. Leslie Boyer who is the director of the VIPER Institute in Tucson, Arizona, "Toxicologists don't really distinguish one molecule as being poisonous and another as benign." "It's all a matter of dose." "So a small dose of something might be a medicine, and a large dose becomes a poison." (Scutti, 2017, p. 3).

Figure 2. One of Dr. Markey's female Wagler vipers

This author's personal experiences working with *Echis* and other venomous species, such as *Elapids* and *Crotolus* as a licensed venomous keeper in the State of Florida and North Carolina revealed that venom toxicity varies not only from one species to another, but it also varies from one individual within that same species to another. It even varies from geographic location to geographic location, as well as differing between the seasons—with a peak during the summer months. Research has suggested that while the toxicity of females' venom is greater on average than males, the males tend to produce a greater volume of venom. The number of factors impacting the composition of the venom potentially injected by each snakebite illustrates the degree of complexity involved with not only the R&D of the antivenom, but also the need for the services of venom banks, such as the one maintained by the Miami Dade Fire Department's special unit also known as *Venom One.* This author maintained an ongoing relationship with these emergency responders for many reasons, such as the access to the antivenom for a wide range of venomous species, the short shelf-life or viability of antivenom, and the major financial investment relating to the purchase and maintenance of antivenom which is often not covered under an individual's health insurance.

Figure 3. Dr. Markey's male Wagler viper (one of her "Wags")

BIOPHILIA: A MEANS TO REMEDIATE PSYCHOLOGICAL SUFFERING

For over 40 years, pet therapy (aka animal-assisted therapy) has been known by several different monikers, but research continues to suggest that there are numerous psychological and emotional benefits experienced across all age groups—for children and the elderly in particular. Pet therapy is utilized in clinical psychology programs to address social and/or emotional issues and communication disorders. It also offers psychological benefits as they relate to stress reduction and/or feelings associated with loneliness and isolation as well as contributing to the individual's quality of life. The theoretical perspective of pet therapy and animal-assisted therapy assumes that chronic stress is a major risk factor in a number of biological and mental health conditions, and animal companionship can aid in mitigating stress levels, as well as promoting relaxation. In turn, this can reduce risk levels and/or symptoms in stress-related conditions.

Chronic feelings associated with isolation, alienation, or marginalization can also exacerbate risk factors for biological and mental health conditions, and animal companionship has indicated a measurable reduction in these feelings. Even the tactile stimulation that is involved when touching, petting, or stroking an animal provides a pleasurable form of neurological stimulation that encourages relaxation and well-being. This stimulation can be mutually beneficial to both parties. The emotional connections formed, and the engagement involved in promoting a relationship with a companion animal, can foster growth—both as to emotional maturity and self-awareness, as well as in promoting a sense of responsibility and loyalty for the companion. Individuals who have been victims of trauma and abuse or who display communication disorders, can perceive non-humans as less threatening than people. As a result, companion animals can raise the children's level of the awareness of their sense of self, as well as serve to encourage a more socially interactive environment among the people around them (Pet Therapy, 2011).

Previous research has suggested that pet therapy has indicated positive outcomes across varying situations and circumstances, as evidenced by the following examples:

1. Hospice settings where pet therapy can benefit both patients and staff, can encourage staff-patient interactions, can improve patient-visitor relationships as well as improve patient morale.
2. Institutionalized settings where pet therapy can reduce depression, lower blood pressure, reduce irritability and agitation, as well as increase social interaction. Support for pet therapy with Alzheimer's dementia patients remains preliminary, but there is evidence that a companion animal's presence can increase social behaviors and the display of emotions, such as smiling, laughing, touching, leaning, verbalizations, etc.
3. Hypertensive patients who take conventional blood pressure medication may have additive value when owning pets.
4. Residents of long-term care facilities—especially those who have a prior history of pet ownership—have shown a reduction in loneliness and depression, as well as an increase in verbal interactions among the residents in the presence of pet therapy.
5. Psychiatric patients have also shown that the presence of a pet dog promotes social interactions. For instance, in schizophrenic patients, there is evidence that pet therapy can lead to an increased interest in rewarding activities, better use of leisure time, and an improvement in motivation, socialization skills, independent living, and overall well-being.
6. Pain management for children based upon preliminary research suggests that canine visitation therapy or CVT can be an effective adjunct therapy in addition to the more traditional pain management modalities for children.

Although dogs and cats tend to be more often associated with pet therapy, fish should not be excluded as an option, since evidence from limited research suggests that pet therapy ala fish aquarium in an institutionalized care facility for Alzheimer's patients can improve nutritional intake and weight gain/maintenance, and reduce the need for nutritional supplements (Pet Therapy, 2011). Naturalistic observations personally conducted by this author in 2014 in North Carolina revealed that both children and adults who have been diagnosed with autism spectrum disorder also achieve significant benefits when interacting with fish and other exotic animal species. The Pets for Vets organization is devoted to the perfect pairing of a United States military veteran with the skilled animal companion that best suits the needs of BOTH the veteran and the companion pet. Pets for Vets has also placed cats and rabbits, and even a pig as consideration for placement in 2014. Pets for Vets employs a unique philosophy that the organization has termed the *Super Bond* (2017), which is based upon the biophilic principles of love, respect, reverence, and collaboration, and which the organization describes as an intense and immediate life-long connection. This unique bond allows the companion animal to immediately begin to offer the veteran the support necessary to ease his or her psychological trauma, mitigate feelings of loneliness, and provide them both with a sense of purpose. Some companion animals have even been observed performing behaviors that they had not been trained to do, such as waking their veterans when they were experiencing nightmares. In the military culture, service members always watch out for each other, knowing that someone always "has their back," and the same is true for veterans and their Pets for Vets companions (Super Bond, 2017).

BIOPHILIA: NOT ALL LIVING THINGS ARE ANIMALS

The Theory of Biophilia embraces the innate relationship that is hardwired into all species who occupy the Earth's biosphere, and to that end, the benefits of plants ranging from their obvious medicinal qualities to their psychological and spiritual influences across cultures should not be understated. A cursory search online indicates the popularity this philosophy has more recently achieved from both an historical, cultural, and spiritual perspective. The Japanese culture refers to it as *shinrin-yoku* (Miazaki, 2018) which translates into the term, *forest bathing*. It developed during the 1980s, and has become the basis of preventive health care and healing in the practice of Japanese medicine. Researchers there have extensively published literature that support their claims to the benefits of spending time in the living forest, which has helped in spreading forest therapy throughout the world. The premise is simple as well as intuitive: Go to a forest and walk slowly. Open all the senses and breathe in the medicine of just being there. The scientifically supported benefits of shinrin-yoku include: an increased functioning of the immune system; a decrease in blood pressure; a reduction in stress levels; an improvement in mood; an increased ability to focus even in populations who have been diagnosed with ADHD; a faster recovery from illness or surgery; an increased energy level; and an improvement in sleep routine. Impressive results were also noted in an individual's:

- Intuitive abilities (deeper and clearer)
- Flow of energy and life force (increased)
- Capacity to communicate with the land and non-human species (improved)
- Friendships and relationships (deepened)
- Feelings of happiness (increased overall)

The establishment of the Association of Nature and Forest Therapy Guides and Program in the United States is an example of just one of the organizations dedicated to the shinrin-yoku therapeutic approach. Their guides have been specially trained to combine leisurely walks under forest canopies with guided activities that serve to open the senses, hone intuitive abilities, and experience the forest as it has never been perceived before. The association initiates individuals into the philosophy of shinrin-yoku so that they can continue the practice independently and reap its benefits. It raises public awareness and mobilizes the innate connection humans have with Nature. It helps to establish areas of natural forests on public lands as designated and protected sites for forest therapy activities including walking and "sit spots" for meditative practices. It helps to transform the relationship between human culture and forests by fostering a collaborative relationship through positive experiences with Nature (Shinrin-yoku, 2019).

THE BENEFITS OF BIOPHILIA IN TIMES PAST

To the Celts, in contrast to modern thought, their integral connection with Nature was what humans call *animism* today. Animism was a belief system that honored the forces of Nature in the form of non-human entities, and in which plants and animals possessed a spiritual essence. They also believed that humans were able to establish a rapport with these spiritual beings. The ancient Celts believed that the supernatural was a central part of Life and that it was interconnected with the material world, so that every tree, spring, marsh river, etc. was spirited. In tribal territories, the earth as well as the waters that received their dead were imbued with sanctity and revered by their living relatives. Sanctuaries were considered to be sacred spaces that were separated from the "ordinary" world, and were designated in natural locations, such as groves, lakes, springs or bogs. The spirits inhabiting watery sites were viewed as the givers of life, and links between the mundane physical realm and the other spiritual world. Even the weather (displayed through its meteorological patterns and phenomena, such as wind, rain, thunder) were viewed as inspirited and propitiated. The Celts believed that their old trees—such as the bile trees—were sacred, and they often served as the social and ceremonial meeting places for a village or tribe.

The animals living within the Celts' environment influenced every area of their everyday life, from their economy, to hunting and waging war, to religious beliefs and rituals, to art and literature. Particular spirits were associated with certain animals, and while some animals were held sacred in their own right, other animals were believed to be messengers from their gods. Boars and deer were frequently represented in the Celtic myths, on statues, on coins, and in literature—since boars appeared to symbolize the characteristics of bravery and prowess in battle, as well as royalty. Bulls and cows were associated with material wealth within the tribe, and were viewed as symbols of the land because meat, leather, milk, and other dairy products were highly valued by the community. Dogs also appeared in many Celtic myths, where they embodied all of the characteristics associated with "a best friend," such as loyalty, companionship, and protection. Since the Celts were an agrarian group, dogs played an important role in hunting, as well as the protection of their flocks. They were also thought to possess great strength as warriors, and ferocity in battle; dogs were often named in association with a "hound." Until recently, dogs' saliva was linked to healing properties, most specifically with the healing of wounds. Since the horse is associated with several Celtic goddesses relating to sovereignty, war, and fertility, it is suggested that the horse also served as a *Psychopomp,* carrying the dead to the otherworld. However, birds (and swans in particular), were thought to serve a more romantic function among the animals. They were a means for the Celtic gods to seek a union with mortals—as evidenced by several well-known myths, such as Leda, and the conception of Castor and Pollux.

It is not uncommon for Celtic wedding bands to be inscribed with a swan representation (Mulvihill, 2019).

BIOPHILIA IN INDIGENOUS THOUGHT

Among the indigenous peoples, conservation is an integral part of their culture, as evidenced by the Maori people of New Zealand. The Maori believe that humans are deeply connected with Nature—to a point where they are considered to be equal and interdependent, even kin. The Maori word, *kaitiakitanga* means "guarding and protecting the environment in order to respect the ancestors and secure the future," which is why marginalized groups such as the Maori are being recognized as stewards of the Earth and the Earth's rapidly disappearing resources. Approximately 370 million people constitute indigenous tribes, which may be only five percent of the world's total population; but they officially hold 18 percent of the land, and lay claim to much more than that. Their geographical home areas occupy 70 countries and extend from the Arctic to the South Pacific, including many of the Earth's biodiversity "hotspots."

Their traditions and core beliefs suggest that they regard Nature with deep respect, and they display a very strong sense of place and belonging—which often runs contrary to the development being imposed from beyond their communities. They fiercely defend their ancestral lands from illegal encroachments as well as destructive exploitation, which includes the construction of mega-dams across their rivers as well as logging and mining operations in their forests. Their efforts can make them ideal custodians and guardians of the landscapes and ecosystems that are relevant to mitigate climate change and to adapt to its effects, but simultaneously it can also make them targets. Indigenous communities who resist powerful economic and political interests can endure intense pressure in many parts of the world. For example, the campaigning group, Global Witness (2015) purported that 185 individuals (many from indigenous tribes) across 16 countries were killed while defending their land, forests, and rivers against industrial destruction in the year 2015 alone.

Finally, after decades of discrimination and marginalization, the role of indigenous peoples as custodians of the land as well as the value of their traditional knowledge underpinning it is being recognized together with their rights to ancestral lands. A report published by the World Resources Institute in 2016 purported that securing the land rights of indigenous tribes and other local communities in the Amazon region was an economical way to counter global deforestation and climate change, which is actually a natural extension of people who just cannot fathom their lives being divorced from Nature (Ranganathan et al., 2016). Conservationists have finally come to the conclusion that the landscapes that they had classified as *wilderness,* have been successfully maintained and protected by local and indigenous communities who possessed the necessary knowledge and skill to manage them. The rights of the indigenous peoples are now reflected in the policies of governments, the strategies of conservation organizations, and within the United Nations Declaration on the Rights of Indigenous Peoples (U.N. Environment Programme, 2017).

CONCLUSION

The *Theory of Biophilia* articulates an unspoken understanding that exists among all of the species that share the Earth's biosphere. The human disruption of the ecosphere suggests a paradigmatic shift from a biophilic perspective (a partnership with Nature) to a biophobic perspective (an adversarial attitude toward Nature). Awe, reverence, and empathy seem to underpin the idea of biophilia, yet many living species continue to be scorned and animal sentience denied. The authors argue that biophilia may offer cures to intractable diseases, and remediate psychological suffering. Biophilia may expand their perspectives, if humans only widen their definition of Life, remember how humans used to view plants and animals, and let indigenous peoples show them how to preserve the native inhabitants of their dying world.

REFERENCES

Allen-Hermanson, S. (2018). Animal consciousness. In *The Routledge Handbook of Consciousness* (pp. 388–407). Taylor and Francis; doi:10.4324/9781315676982-30

American Pet Products Association. (2015). *APPA National Pet Owners Survey 2015-2016*. Greenwich, CT: American Pet Products Association.

Awe. (2019). In *Merriam-Webster.* Retrieved from https://www.merriam-webster.com/dictionary/awe

Bariso, J. (2019). There are actually 3 types of empathy. Here's how you can develop them all. *Inc.* Retrieved from https://www.inc.com/justin-bariso/there-are-actually-3-types-of-empathy-heres-how-they-differ-and-how-you-can-develop-them-all.html

Biophilia. (2019). *Cambridge Dictionary.* Retrieved from https://dictionary.cambridge.org/us/dictionary/english/biophilia

Biophilia. (2020). *Dictionary.com.* Retrieved from https://www.dictionary.com/browse/biophilia

Birch, J. (2019). *Seven distinct Greek words describe different kinds of love—which have you experienced?* Retrieved on March 1, 2020 from https://www.wellandgood.com/good-advice/greek-words-for-love/

Buss, D. M., & Barnes, M. (1986). Preferences in human mate selection. *Journal of Personality and Social Psychology, 50*(3), 559–570. doi:10.1037/0022-3514.50.3.559

Carson, J., & Griffith, S. (2009, September 2). *What is the for-profit social enterprise?* Yale Insights. Retrieved on March 14, 2020 from https://insights.som.yale.edu/insights/what-is-the-for-profit-social-enterprise

Chalmers, D. J. (1995). Facing up to the problem of consciousness. *Journal of Consciousness Studies, 2*(3), 200–219. Retrieved from http://consc.net/papers/facing.html

Danchin, E., Nobel, S., Pocheville, A., Dagaeff, A. C., Demay, L., Alphand, M., ... Isable, G. (2018). Cultural Flies: Conformist Social Learning in Fruitflies Predicts Long-lasting Mate-choice Traditions. *Science, 362*(6418), 1025–1030. doi:10.1126cience.aat1590 PMID:30498121

Empathy. (2019). *Psychology Today.* Retrieved on February 27, 2019, from https://www.psychologytoday.com/us/basics/empathy

Freud, S. (1917). Mourning and melancholia. In The Standard Edition of the Complete Psychological Works of Sigmund Freud, Volume XIV (1914-1916): On the History of the Psycho-Analytic Movement, Papers on Metapsychology and Other Works, (pp. 237-258). London, UK: Hogarth Press.

Freud, S., & Brill, A. A. (1913). *Totem and taboo: Resemblances between the psychic lives of savages and neurotics*. London: G. Routledge.

Fromm, E. (1964). *Heart of man: Its genius for good and evil.* New York: Harper & Row.

Fry, B. G. (2019). *Venom Doc.* Venom Evolution Lab. Retrieved on March 3, 2019 from http://www.venomdoc.com/support-our-research

Gerber, T. (2019, February 16). Eros and Thanatos: Freud's two fundamental drives. *Epoché.* Retrieved on March 14 from https://epochemagazine.org/eros-and-thanatos-freuds-two-fundamental-drives-50a82a11a389

Gilhus, I. S. (2006). *Animals, gods, and humans: Changing attitudes to animals in Greek, Roman, and early Christian thought.* London, UK: Routledge. doi:10.4324/9780203964798

Global Witness. (2015). *Guns, cronies and crops: How military, political and business cronies conspired to grab land in Myanmar.* Global Witness.

Gould, S. J. (1994). Scales of destruction. *The Sciences, 34*(3), 36–38.

Grandin, T. (2005). How animals perceive the world. In T. Grandin & C. Johnsons (Eds.), *Animals in translation: Using the mysteries of autism to decode animal behavior* (pp. 27–67). New York, NY: Houghton Mifflin Harcourt.

Henderson, W. (2017). A brief history of hemophilia treatment. *Hemophilia News Today.* Retrieved on March 4, 2019, from https://hemophilianewstoday.com/2017/05/15/brief-history-hemophilia-treatment/

Hodson, G., Kteily, N. S., & Hoffarth, M. R. (2014). Of filthy pigs and subhuman mongrels: Dehumanization, disgust, and intergroup prejudice. *TPM. Testing, Psychometrics, Methodology in Applied Psychology, 21*(3), 267–284. doi:10.4473/TPM21.3.3

Jahr, G. H. G. (1857). General and special therapeutics of mental diseases and psychical disorders (J. M. Galloway, Trans.). Manchester, UK: Kessinger Publishing.

Joye, Y., & De Block, A. (2011). Nature and I are two: A critical examination of the biophilia hypothesis. *Environmental Values, 20*(2), 189–215. doi:10.3197/096327111X12997574391724

Jylhä, K. M., Cantal, C., Akrami, N., & Milfont, T. L. (2016). Denial of anthropogenic climate change: Social dominance orientation helps explain the conservative male effect in Brazil and Sweden. *Personality and Individual Differences*, *98*, 184–187. doi:10.1016/j.paid.2016.04.020

Maizels, N. (in press). *The life instinct.* Retrieved from https://www.researchgate.net/project/The-Life-Instinct

Marchant, J. (1646). *Rationale evangelizantium, sive doctrina et veritas evangelica, a sacerdotibus, pastoribus, concionatoribus, pectori appendenda, plebique per anni circulum è Cathedris proponenda, exponenda: Duobus tomis comprehensa.* Paris, France: Apud Michaelem Soly.

Meinecke, L. D. (2017). *Neglected by assessment: Industry versus inferiority in the competition for scarce kidneys* (Doctoral dissertation). Available from ProQuest Dissertations and Theses database. (ProQuest No. 10689852)

Meinecke, L. D. (2018a). A theory of cognitive idolatry, or the human struggle to preserve psychological certainty in a biological species - chapter 1 (Research proposal). doi:10.13140/RG.2.2.16527.43684

Meinecke, L. D. (2018b). A theory of cognitive idolatry, or the human struggle to preserve psychological certainty in a biological species - chapter 2 (Research proposal). doi:10.13140/RG.2.2.31089.53601

Mendl, M., & Paul, E. S. (2008). Do animals live in the present? Current evidence and implications for welfare. *Applied Animal Behaviour Science*, *113*(4), 357–382. doi:10.1016/j.applanim.2008.01.013

Meyer, M., & Bergel, K. (2012). *Reverence for life: The ethics of Albert Schweitzer for the Twenty-First Century.* Syracuse, NY: Syracuse University Press.

Miazaki, Y. (2018). *Forest Bathing: The Japanese Way of Forest Bathing for Health and Relaxation.* London, UK: Aster, a Division of Octopus Publishing Group, Ltd.

Morell, V. (2013). *Animal Wise: The thoughts and emotions of our fellow creatures* [Kindle Edition]. New York: Crown.

Mulvihill, C. (2019). How nature shaped Celtic culture in Ireland. *Green News Ireland.* Retrieved on March 4, 2019, from https://greennews.ie/how-nature-shaped-celtic-culture- in-ireland/

Murphy-Gill, M., & Johnson, S. (2016, May 20). Do dogs go to heaven? Animals are part of God's creations, but will they join us in heaven? *U.S. Catholic*, *81*(6), 49. http://www.uscatholic.org/articles/201605/do-dogs-go-heaven-30654

Nordstrom, C. (2004). *Shadows of war: Violence, power, and international profiteering in the twenty-first century*. Berkeley, CA: University of California Press.

O'Grady, M. A. (2016). *The theory of biophilia* (Reg. No. TXu 2-032-922). United States Copyright Office. Retrieved from https://www.copyright.gov/

Perseus Digital Library. (2020). *Greek word study tool*. Retrieved from http://www.perseus.tufts.edu/

Pet Therapy. (2011). Retrieved on March 4, 2019, from http://www.naturesmarketholland.com/ns/DisplayMonograph.asp?storeID=ac1be0a89b594ef78157febc945dabfe&DocID=bottomline-pettherapy

Radkowsky, M. (2019). Research with nonhuman animals. *Monitor on Psychology*, *3*, 7.

Ramachandran, V. S., & Hubbard, E. M. (2001). Synaesthesia: A window into perception, thought and language. *Journal of Consciousness Studies*, *8*(12), 3–34. Retrieved from http://cbc.ucsd.edu/

Ranganathan, J., Vennard, D., Waite, R., Dumas, P., Lipinski, B., Searchinger, T. I. M., & GLOBAGRI-WRR, M. A. (2016). Shifting diets for a sustainable food future. *World Resources Institute*. Retrieved on March 1, 2020 from http://www.indiaenvironmentportal.org.in/files/file/Shifting_Diets_for_a_Sustainable_Food_Future.pdf

Rapaport, S. I., Aas, K., & Owren, P. A. (1954). The clotting action of Russell viper venom. *Blood*, *9*(12), 1185–1194. doi:10.1182/blood.V9.12.1185.1185 PMID:13208768

Respect. (2019). In *Merriam-Webster*. Retrieved from https://www.merriam-webster.com/dictionary/respect

Reverence. (2019). In *Merriam-Webster*. Retrieved https://www.merriam-webster.com/dictionary/reverence

Rodgers, R. (1959). Do-re-mi [Recorded by J. Andrews]. On The sound of music [Vinyl record]. Los Angeles, CA: RCA Victor.

Rogalski, A., Soerensen, C., Op den Brouw, B., Lister, C., Dashevsky, D., Arbuckle, K., ... Wüster, W. (2017). Differential procoagulant effects of saw-scaled viper (Serpentes: Viperidae: Echis) snake venoms on human plasma and the narrow taxonomic ranges of antivenom efficacies. *Toxicology Letters*, *280*, 159–170. doi:10.1016/j.toxlet.2017.08.020 PMID:28847519

Safina, C. (2015). *Beyond words: what animals think and feel*. New York, NY: Picador.

Scutti, S. (2017). Snake venom is key ingredient in experimental drug for heart patients. *CNN.* Retrieved on March 3, 2019, from https://www.cnn.com/2017/06/08/health/snake-venom-antiplatelet-drug-study/index.html

Shinrin-yoku. (2019). *Shinrin-Yoku.org*. Retrieved on March 3, 2019, from http://www.shinrin-yoku.org/shinrin-yoku.html

Skinner, B. F. (1957). *Verbal behavior*. Englewood Cliffs, NJ: Prentice-Hall. doi:10.1037/11256-000

Spinoza, B., & Curley, E. M. (1994). *A Spinoza reader: The Ethics and other works*. Princeton, N.J: Princeton University Press.

Super Bond. (2017). Pets for vets. *The Super Bond™: Creating an Unbreakable Connection.* Retrieved on March 4, 2019, from https://www.petsforvets.com/super-bond

Suskin, S. (2009). *The sound of Broadway music: a book of orchestrators*. New York, NY: Oxford University Press, Inc.

Tuke, D. H. (1892). *A dictionary of psychological medicine*. London, England: J. & A. Churchill.

U.N. Environment Programme. (2017, April 26). *Indigenous people and nature: A tradition of conservation*. Retrieved on March 14, 2020 from https://www.unenvironment.org/news-and-stories/story/indigenous-people-and-nature-tradition-conservation

Vatican News. (2019, May 27). *Pope to Caritas Internationalis: The meaning of charity*. Retrieved on March 14, 2020 from https://www.vaticannews.va/en/pope/news/2019-05/pope-to-caritas-internationalis-the-meaning-of-charity.htm

Weir, P. (2006, May 30). Green about the genes. *The Guardian*. Retrieved from https://www.theguardian.com/education/2006/may/30/workinprogress.highereducation

Wilson, E. O. (2016). *Half-earth: Our planet's fight for life*. New York, NY: Liveright Publishing.

Wise, R. (Producer/Director). (1965). The sound of music. Beverly Hills, CA: Twentieth Century Fox Home Entertainment.

ENDNOTES

1 Freud called these opposing wishes *Eros* (the life instinct or libidinal impulses) and *Thanatos* (the death instinct or destructive impulses). His "disciple" Fromm called them *biophilia* (love of life) and *necrophilia* (love of death).

2 Evo-devo: an evolutionary/developmental biological lens.

3 Silent partner: a party whose role in a partnership is limited to providing working capital to the business.

4 SDO: Social Dominance Orientation.

Chapter 2
Beyond Biophilia:
The Theory of Astrophilia

ABSTRACT

Chapter 2 expands the concept of biophilia to encompass an awe for the majesty of creation beyond the Earth's biosphere. Also treated in this chapter is the need for a new ethics that is more inclusive of the unknown, much as the Theory of Biophilia calls for more inclusion of the unrepresented. There will be a need to redefine awe and rethink companionship, as human explorers endure extended time away from reminders of home in the pursuit of endless discovery. This chapter will take the reader to where no theory has gone before.

INTRODUCTION

Astrophilia is the theory that takes over where the Theory of Biophilia leaves off – in other words, Astrophilia may be defined as "biophilia beyond the borders of our biosphere." The Theory of Astrophilia, as it was submitted to the United States Copyright Office on January 1, 2019 (Markey, 2019) is rather mind-bending at first glance because it ventures unto the under-charted or lesser-charted territory of Space. It reads as follows:

The Theory of Astrophilia suggests that sentience as defined as a capacity to feel, perceive or experience subjectively, combined with awareness as defined as a recognition of what we are not, provide the means through which we

DOI: 10.4018/978-1-7998-4408-2.ch002

are inextricably linked with, and innately motivated to seek out the "other" with whom we share the Universe (Markey, 2019).

The reader will recall that the *Theory of Biophilia* addresses the innate relationship that exists among all the species that share the biosphere on Earth. Since the idea of awe is not confined to the Earth's biosphere, the Theory of Astrophilia proposes that a similar relationship exists among all sentient beings that share a common universe. Even beyond the familiar spheres confined to the gravity of the human species' own world there lies a celestial sphere, one might say, with which all beings are inextricably linked, be they confined to gravity, or be they beyond its influence. Thus, a logical extension of the prior arguments of collaboration among the Earth's kinds suggests that the same collaboration is vital far beyond current understanding.

Figure 1. The logo for the author's Theory of Astrophlia™

CONSTRUCTING A NEW ETHICS: OUR PRIME DIRECTIVE

New ways of thinking about awe require new ways of thinking about environmental protection, and whether the societal indifference to the environment on Earth might accidentally translate to a similar indifference out in space. As you recall, the Theory of Biophilia posits that, when humans motives suggest they are undergirded by more noble ideas such as awe, reverence, respect, and empathy, both humans and whatever humans come into contact with have the ability to flourish. But when humans motives are ignoble, or are guided primarily by egocentrism, self-serving biases, devaluation, and the need to dominate whatever humans discover, the end result of that collision with the great unknown is more likely to be catastrophic for us both. Nowhere is this more frightening than the idea of spreading human pollution and an insatiable greed beyond this dying, modest biosphere.

As you can see, the Theory of Astrophilia seems to suggest the need for a philosophy that is similar to the *Prime Directive* introduced by the science fiction writer, Gene Roddenberry (1921-1991), in his *Star Trek* series (1966-1969). In essence, it was the prohibition of interference with the natural development of other cultures and other civilizations by representatives of Starfleet during the crews' exploration of the Universe. Starfleet Order One was particularly aimed at the prevention of interference with a civilization's internal development if they should happen to be less technologically advanced. It can be summarized as follows:

- No identification of self or mission
- No interference with the social development of said planet
- No references to space, other worlds, or advanced civilizations

However, even though it was drafted by Roddenberry with the best of intentions, many viewers noticed that the Prime Directive seemed to be violated with a certain degree of regularity during several of the series' episodes, when a strict adherence to it was not desirable or required (Stemwedel, 2015).

Examining the Prime Directive from an ethical perspective, its wording is not unlike the wording humans use in human subjects research, in order to guarantee the protection of sentient beings, and afford the uninformed the same treatment that scientists would expect to receive from each other. In brief, human subjects research is based on the Belmont Report, and in that report are outlined three fundamental principles:

- Autonomy
- Beneficence
- Justice

The principle of *autonomy* is a regard for persons which cannot be diminished by the objectives of scientific research or exploration (Borenstein, 2017). Within the Earth's biosphere, the Theory of Biophilia expands this definition so that it does not exclude individuals simply because they do not appear to be human or motivated by human aims. In so doing, it seeks to guarantee protections for plant and animal species with which humans share their biosphere, and without which Life on this planet would not be possible.

Out in space, the Theory of Astrophilia would challenge the definition of Life and seek to expand that definition so as not to exclude the idea of a sentience dissimilar to that of the average human being.

The principle of *beneficence* in human subjects research is about upholding the well-being of those from whose participation researchers benefit during their research. On the Earth, the concept of biophilia embraces the need to safeguard the well-being of non-human animals, of lakes and skies and grasses whose fragile ecosystems and inalienable need for ecological balance are not unlike the regulatory processes inside the humble universe of a single human organism. Out in the cosmos, the Theory of Astrophilia posits that this same fragility continues, and therefore the same need for beneficence is required.

Finally, the principle of justice in scientific research involving human subjects remains concerned with protecting the activities and proceeds of research from injustice, moreso than those to whom the principle of justice ought to apply (be they human or non-human in nature). The end result of such a perspective suggests that when the needs of the many outweigh the needs of the few, the few can be denied justice (or vice versa). This theme was actually prominent in one of the Star Trek Films, *Star Trek II: The Wrath of Khan* (Sallin & Meyer, 1982). In true Roddenberry fashion, however, the principle of mercy for Life and the unswerving loyalty underpinning human and non-human friendship outweighed every other principle—the needs of the few outweighed the needs of the many, because mercy apparently is not constrained by the quantity of persons needed to unconditionally dispense mercy. Nowhere is this demonstrated more poignantly than in the proposition that any benefits realized by the human exploration of space should not exceed any benefits realized by the beings humans might encounter in the course of that exploration. In other words, human exploration is done only for its own sake, and not for gain or profit.The Prime Directive, like the Theory of Astrophilia, reflects a consequentialist dedication to mitigating harm, as well as a Kantian commitment to respecting others' right to autonomy—since the assumption that other cultures fare better when left to their own devices has been incorporated into it. The term "devices" may be defined as a civilization's social practices or their concrete technologies. Interference by a member of Starfleet even when motivated by the best of intentions was likely to complicate things in unanticipated ways, so that even when a culture had to cope with unintentional harm, that culture would experience less anticipatable harm if those intentions originated from that culture's own free will. This philosophy embodies a kind of anti-colonialist ethos by respecting the values, beliefs and practices of other civilizations instead of imposing a "better" set of core values, etc. upon unwitting cultures. Adherence to a non-interference policy reflects a certain type of attitude toward the project of investigating other cultures by employing a strategy known as *disguised naturalistic observation* in which the observers inconspicuously watch, but do not interact with, the target subjects within a civilization. The goal of this strategy is to provide the disguised observers with the opportunity to witness how the culture would behave if the observers were not there, since any type of intervention is presumed to influence or contaminate "natural" behaviors as well as the eventual development of indigenous technologies (Stemwedel, 2015). One caveat to this struggle to observe without influencing the outcome, is that observation (at the quantum level) always influences the outcome.

The Prime Directive also aligns with strong moral intuitions by suggesting how vital it is to respect the autonomy of other *cultures* (not just other individuals) that are not of this Earth, as well as to "do no harm" by avoiding the possibility of inflicting unintentional harm. If this idea should apply to indigenous and non-human cultures on this world, it should apply to indigenous and non-human cultures throughout the cosmos, wherein an unknowably indefeasible autonomy existed long before the Earth was formed.

So, in summary, such an ethics—on this homeworld and beyond—promotes a more *universal* ethics, one of *sharing* the Universe and its majesty, rather than trying to *claim* or conquer it in the name of special interests that have the power to do so. By sharing the Universe with other beings, any thoughts of domination or colonization are removed "from the table" as the attitude becomes one of collaboration, and if not friendship, at least reverent tolerance for one another. Imagine if this ethos, not human expansionism, had been the prime directive during the colonization of the New World? Since not all situations present themselves as being "black and white" even in Space, circumstances may arise where choosing *not* to intervene could result in harming a civilization more greatly than the alternative intervention, such as sharing a technology that could prevent the extinction of that civilization. Perhaps one solution to such a dilemma might be to allow the civilization in distress to make the choice as to whether they would accept the offer of the shared technology or not. There is also the possibility that efforts to respect the autonomy of a civilization could be perceived as *paternalistic* motives by withholding pertinent information about self, mission and/or other worlds (e.g. the knowledge that one is being observed). There is a reciprocity involved with sharing the Universe with others as well as the granting of mutual respect and moral consideration even when the level of technological attainment is different. Perhaps the Prime Directive should not be considered as "exceptionless" after all, since ethics may require establishing a sense of trust in other civilizations' abilities to make their own decisions after making *first contact* (Stemwedel, 2015).

REDEFINING AWE BY CHALLENGING THE DEFINITION OF LIFE

This author has taken great care in avoiding (as much as possible) the term, *life,* as it relates to Space since up to this point in time, scientists have been grappling quite a while with constructing a scientific definition of what constitutes life here on Earth, and yet a just and accurate meaning seems to elude definition. For this reason, when addressing the topic of Biophilia in Space or "Biophilia beyond our biosphere," the conundrum becomes even worse. According to NASA—whom one would think would have easily mastered this semantic challenge with some degree of certainty by now—constructing a working definition of this term (life) is complicated by divergent goals for its definition by diverging fields of science. These include biology, biochemistry and genetics, to name a few. NASA compares it to the Hindu tale of attempting to identify an elephant by inviting six blind men to study it, and by having each study a different part (the tail, the trunk, the leg, etc.). Though each may comprehend a piece, none of them comprehends the organism. This suggests that each scientific discipline would ultimately arrive at a different definition by viewing the problem only through his or her specific lens or viewpoint.

Surprisingly however, some initial agreement has been possible with regard to the identification of multiple characteristics that are typically affiliated with life. "Living things" tend to be very complex and highly organized; in addition, they have the ability to absorb energy from their environment and convert it for growth and replication or reproduction. Simple life forms prefer to remain in *homeostasis* or a state of equilibrium that defines their internal state, and they respond to stimuli which fosters a reaction type of motion and recoil (think Newton's Laws of Motion), while advanced forms of life learn as well (Glenn Research Center, 2015). Living things reproduce and multiply since some type of copying mechanism is essential for evolution to occur via natural selection and mutation. In addition, living things must be consumers in order to grow and develop—which by definition often includes changes in biomass, the creation of new individuals, and the disposal of waste material. So, in order to be classified as a living thing, an individual must conform to these criteria. One familiar example to consider, however, is a crystal—which has the ability to grow, reach equilibrium, and react or move in response to stimuli, but is lacking a biological nervous system. While a "bright line" definition is highly desirable, borderline cases that baseline human beliefs cannot handle continue to blur the definition into a rather grey and fuzzy configuration.

In the hope of more accurately operationalizing the definition terrestrially, all known organisms appear to share a chemistry that is 1) carbon-based, 2) demonstrates a dependence on water, and 3) leaves behind fossil remains containing carbon or sulfur isotopes that indicate a present or a past metabolism. Terrestrial life has been classified into four biological families: *archaea, bacteria* or *prokaryotes, eukaryotes*, and *virus*es where archaea or extremophiles are the most recently defined branch. Archaea typically flourish in extreme environments as single-celled beings but also share characteristics with bacteria and eukaryotes. Bacteria usually lack chlorophyll (with the ever-present rule-breaker exception, cyanobacteria) and a cell nucleus, and ferment and respire to generate energy. The eukaryotes include all organisms with cells that have a nucleus, such as humans, animals, plants, protists, and fungi. The final group includes viruses that contain fragments of DNA and RNA which reproduce parasitically when they infect a suitable host cell. However, while these four classifications help to clarify some of the questions underlying what constitutes life, they do little to provide a succinct definition which may be compared to past efforts to define *water* prior to the existence of *molecular theory* (Watanabe, 2007). As well, the simple establishment of fundamental categorical expectations, as the reader will see in chapter three, is prone to paradox.

Now attempt to extend that elusive definition beyond the biosphere of the Earth, and the situation assumes an even greater challenge—particularly with the realization that until this point in time *Life* has been defined by what is known, by what has been seen previously, and by what has been cultured in a laboratory petri dish. In an interview, Carol Cleland who co-authored the paper entitled, *Origins of Life and Evolution of the Biosphere* with Christopher Chyba, argued that it is a mistake to attempt to define "life" because such efforts illustrate the fundamental misunderstandings that pertain to nature and the power of definitions (Cleland & Chyba, 2002). According to Cleland, definitions communicate the meanings of words within the confines of human languages, in contrast to talking about the nature of the world—which is not confined to the limitations of human languages (for instance, nonverbal behavior). Scientists are not interested in what the word "life" means in human language; instead, scientists focus on constructing a general theory of living systems which extends beyond a single case with a common origin, such as terrestrial life on Earth (despite its morphological diversity). The key for constructing such a general theory of living systems is to investigate alternative possibilities for life, which includes moving beyond an Earth-centric perspective, and formulating a strategy for seeking extraterrestrial life. Cleland suggests that life could have arisen more than one time from non-living materials somewhere other than the Earth or, it could have arisen right here on Earth. It is also possible that extraterrestrial life exists, but that all life shares a common ancestor. Scientists have concluded that microbes are able to survive interplanetary travel while ensconced in meteors that were generated by asteroid collisions with planetary bodies containing life. In other words, humans could all be descendants of Martians, or Martians (should they exist) could share a common ancestor with humans—thereby suggesting that the discovery of extraterrestrial life does not mean that life originated from multiple sources (Watanabe, 2007). So in the future, it may be revealed that the chunk of meteorite on display at the local museum contains human beings' distant relatives "from a galaxy far, far away."

This author's attempt to avoid the minefield plagued by definitions that do not accurately define what they ascribe to articulate, as well as semantics that serve only to obscure meaning, has resulted in a theory. The author had to overcome concepts such as *formal semantics*, which focuses on the logical aspects of meaning (sense, reference, implication and logical form); the author had to circumvent *lexical semantics*, which addresses the meanings and relationship of words; and she had to tackle the subbranch known as *conceptual semantics*, which investigates the cognitive structure of the meaning of words (Semantics, 2019). Despite the foregoing, a theory has coalesced as it relates to the new field of Astrophilia, and is quoted again here:

The Theory of Astrophilia suggests that sentience as defined as a capacity to feel, perceive or experience subjectively, combined with awareness as defined as a recognition of what we are not, provide the means through which we are inextricably linked with, and innately motivated to seek out the "other" with whom we share the Universe (Markey, 2019).

The idea of Life may evade definition, especially outside those spheres humans are accustomed to; still, since the idea of Awe continues just outside the Earth's atmosphere, so too it follows that something like Life itself continues there as well.

ASTROPHILIA: THE CALL FOR BIOPHILIA ENROUTE TO THE STARS

As purported by the Theory of Biophilia (O'Grady, 2016) in chapter one, an undeniable and unescapable connection exists among all species who share the Earth's biosphere, so it is not at all surprising that humans are more comfortable traveling in space when they are accompanied by plants both nutritive and non-nutritive. Scott Bates, the interim department head at Utah State University and his Russian counterpart, Vladimir Gushin at the Russian Academy of Science's Institute for Bio Medical Problems in Moscow, collected and analyzed survey data from astronauts and cosmonauts who had flown and worked with plant experiments in orbit. They also collected and analyzed secondary sources of journal data from the crew of a Russian ground-based simulation of a mission to Mars that was called *Mars500* in 2010 (Bates et al., 2010). Following the receipt of Shane's email, this author was on her own mission to find out more about the motives and the methods for growing plants in Space.

A quote by Bruce Bugbee (yes that's his real name) caught the attention of this author (who is also a pilot), as she began her biophilic journey into Space. In an article about the Space Dynamics Lab, published in 2009, she read: "Are there people that just don't need plants around, and if there are, are those the kind of people we want to be flying? Are those the most mentally stable people the ones that like plants and like interacting with plants?" (Quinn, 2009, para. 3). It soon became apparent that even NASA had realized that plants are a vital part of the Earth's eco-system, and play a crucial role beyond Earth by providing nutritive (food) and psychological (calm and tranquility) benefits to travelers in Space. In Sheri Quinn's 2009 article, Bruce Bugbee was reported as saying that researchers were attempting to quantify the importance of plants but not only in a dietary sense of fresh greens for the crew. Bugbee revealed that since 2002, the International Space Station aka ISS has maintained a small greenhouse garden named *Lada* after the ancient Russian goddess of spring. Lada was constructed to provide *green space* for crew members during the long flights and extended stays in Space, and the suitcase-sized plot produced a steady supply of fresh produce in the form of (predominately) peas and *mizuna,* a fast-growing variety of lettuce. In addition, Bugbee explained that Lada also provided a welcome distraction for crewmembers—as evidenced by an incident he recounted that occurred several years ago when the space shuttle broke up upon re-entry which caused the crew members still living on the space station to become very anxious. In response, the Russian space program assigned more gardening time to each crew member, because they had noticed that tending the plants had a calming effect. So why not employ the idea of "plants in Space" as a psychological tool, to mitigate an astronaut's worry, anxiety, etc. independent of the obvious food value?

Vladimir Gushin, a psychologist at Russia's Institute for Biomedical Problems suggested that investigating the correlation between the amount of time crew members spend tending the garden and the psychological benefits has not generated sufficient data to draw any scientific conclusions. He also goes on to say that confinement on the space station is not problematic ... instead, it is the lack of natural stimuli, such as the blowing of the wind, the chirping of the birds, and other sensory input often taken for granted on Earth, which creates issues for the crew members. Due to cargo limitations on the ship, certain challenges arise when attempting to keep crew members physically and mentally "alive." In Gushin's words, "Plants are one of the opportunities that makes them [the crew members] feel something is changing, that nature is with them, a piece of earth is with them . . . that gives them the feeling that there is still a piece of earth, of life. From this point, nothing can substitute for plants" (Quinn, 2009, p. 3). His quote reflects the essence of the Theory of Biophilia (O'Grady, 2016), which states that all species are "inextricably linked."

Life in Space can be a bit tricky for plants as well as humans, since successful growth requires convincing the plants that they are growing on Earth in natural sunlight. In Lada, plants are taught to grow upwards toward common household fluorescent bulbs, and instead of soil, they are planted in a bed of baked clay particles. Water has to be carefully measured and replaced, since its excess will not drain away in micro-gravity, and air has to be recycled and filtered to remove any trace contaminants that may be toxic to the plants. Gardening in Space for its psychological and nutritive benefits, is not the only gain from Lada's creation and maintenance. Researchers are also learning about the effects of air quality and agriculture on Earth; however, their ultimate goal is to grow food successfully in Space for extended missions destined for interplanetary travel (much as ancient mariners sought a means to conduct distant voyages—on seas where waypoints for resupplying the ocean-going craft and its crew were few and far between). Bruce Bugbee commented that he is looking forward to erecting a greenhouse on Mars, which is not only possible but essential—since it is unlikely that the crew members will be able to transport enough food with them. In addition, he mentions the economic feasibility of moving the entire space farm, solar cells, and plant material, in contrast to transporting a trailer-load of bag lunches to Mars (Quinn, 2009).

PLANTS AS HUMAN COMPANIONS

The concept that plants contribute to the survival of humans as well as non-humans along the food-chain, by providing them with nutrition and a life-sustaining atmosphere, is more well-known and accepted than is the idea that plants provide significant non-nutritive value too. Researchers Bates et al. (2010) outlined a variety of ways in which plants have been indicated to have positive (non-nutritive) influences on humans, and therefore can serve as a countermeasure for problems faced by humans who are living in isolated and/or extreme environments for long durations of time. During the early years of human space travel and limited exploration, the influence of psychosocial facts on crew members' behavior and performance was deemed to be minimal. During short space missions, they were able to cope with stress, tension and interpersonal conflicts without adversely affecting mission goals. (They knew they would not be gone for long).

However, with the advent of space stations such as Skylab, Salyut, Mir, and the International Space Station, also came a significant increase in a crew member's duration out in Space (six months or longer). The composition of crews changed also, becoming both more multinational and more heterogenous with regard to sex, culture, and professional training. Such diversity in the crew's psychosocial and physiological factors—in combination with isolation, monotony, confinement, and a hostile Space environment—exacerbates the stressors native to space travel, and impacts crew performance, morale, and overall well-being (Gazenko, Myasnikov, & Uskov, 1976). Among the observed psychological and behavioral reactions were lapses in attention, sleeping issues, irritability toward other crew members or mission control staff, as well as a considerable decline in motivation and energy (Kanas et al.,, 2001; Kozerenko, Gushin, & Sled, 1999). Most of these reactions are non-specific to Space explorers and are well-known to occur in other isolated and confined environments also. However, in Outer Space (unlike an Earth-bound environment) such negative responses are likely to have disastrous results, when or if a depressed crew member should become unable to perform a critical task during an emergency situation.

In an effort to mitigate the negative effects of isolation, sensory deprivation, and monotony, space psychological support groups (they really exist) focus on providing a variety of sensory stimuli and novel experiences (Grigoriev, Kozerenko, Myasnikov, & Egorov, 1988). Among those activities have been surprise gifts, natural food deliveries via re-supply vehicles, and increased music and lighting onboard the crew members' space facility. Teleconferences with families and friends on Earth as well as access to personal computers containing photos and videos of home, and pleasant places or *favorite landscapes*, have also served a similar purpose, when individually tailored to each crew member. In addition, to prevent the onset of mental disorders, crew members receive counseling or psychotherapy from the ground crew while deployed in Space, as well as medications and sedatives as necessary (Kanas, 1998). While long-duration stays aboard an orbital space station pose a certain set of challenges for crew members, larger endeavors, such as constructing a permanently crewed human outpost on the moon or Mars may create conditions in which psychological problems could create even greater significant challenges for humans (Kanas & Manzey, 2003). The rigors of traveling extremely long distances pose novel challenges. There are long delays in communication and vital human feedback (such as a lack of opportunities for counseling and family interaction via teleconferencing). The uncertainty arising from long waiting periods, either in hope of rescue or in hope of re-supply, will require that future crew members exhibit a much higher level of autonomy and self-confidence. This would be a logical expectation of being exposed to a much greater level of confinement and isolation than any previous space crew or expedition crew on Earth.

MY BEST FRIEND, A BEGONIA

Humans and non-humans have become accustomed to relying on plants for their survival in many ways, both as a source of food and for happily creating a life-sustaining atmosphere on Earth; however, little research had been conducted on the non-nutritive benefits of the human-plant relationship. A literature review on the nutritive benefits of plants revealed an approximate 35-year history which is not generalizable to the conditions or the people living in the extreme evo-psycho-social conditions of Space. However, the literature does suggest that plants could provide positive benefits in extreme conditions or environments. Human-plant interaction varies in the level of *intensity* which is an important consideration when considering the use of plants as a possible countermeasure to the negative effects of isolation, sensory deprivation, and monotony. Low-level interaction can be defined as just looking at a plant or an image of plant life or looking at a plant through a window—in contrast to the more intense interaction of gardening, tending, harvesting, etc. where there is physical contact with the plant. The benefits of exposure to plants have been investigated across several disciplines, and a variety of benefits have been assessed. For example, humans appear to prefer looking at natural scenery as opposed to looking at urban landscapes (Kaplan, 1987; Ulrich, 1986). One particularly relevant study was conducted by Wise and Rosenberg (1988) who assessed physiological responses along with analyzing self-reported preferences to natural scenery within a simulated space station environment where a bulkhead was created to show one of three scenes: a savannah, a mountain scene, or modern abstract art (or blank control scene). The results indicated that the mountain scene was the most preferred, but both nature scenes (the savannah and the mountains) were most effective in impacting the viewers' physiology.

Research conducted by Heerwagen and Orians in 1986 suggested that humans will seek out or even create contact with Nature, when real contact is impossible. This is apparent in the efforts which office workers exhibit in decorating their workspaces. The researchers hypothesized that office workers with access to a nature view as provided by a window would decorate their office spaces differently from office workers who had no such view, and their hypothesis was supported. While all of the office workers utilized more nature-oriented materials as compared with non-nature-oriented materials for decoration, the office workers with windowless offices used *three times more* nature-oriented material than windowed offices. Furthermore, Ulrich (1984) discovered that humans not only prefer to *look* at nature over urban environments, but they also *benefit* from looking at nature. He also found that surgical patients who were assigned to rooms with windows that overlook nature had shorter hospital stays, ingested fewer analgesics, and received fewer negative evaluations from hospital staff than surgical patients who were assigned to rooms overlooking a brick building. These results held true even when other factors, such as age, sex, weight, floor-level, and general health were controlled. More recently in 2006, Fjed stated that office workers who were randomly assigned to plant versus no-plant experimental conditions had fewer health-related symptoms over time.

The viewing of natural scenery as well as the presence of living plants has also been linked to the reduction of stress and stress recovery, which was supported by two studies involving Roger Ulrich, first in 1979, and again with his colleagues in 1991. The first study used college students to examine the impact that images of natural scenery had on levels of stress which were related to a course examination (Ulrich, 1979). The results showed that students viewing images of natural settings actually experienced reduced stress levels relating to the exam. Later in 1991, Ulrich and his colleagues conducted a research study to investigate the effects of video clips showing natural or urban settings, followed by presentation of a psychologically stressful black-and-white video segment lasting 10 minutes (used by industry to reduce accidents on the job; Ulrich et al., 1991). The video clip portrayed a series of graphically gory accidents including mutilations and bloody situations which was immediately following by the viewing of one of six 10-minute color videos of either natural scenes or urban environments. Physiological assessments were then taken including participants' blood pressure, skin conductance, and muscle tension to determine the level of their stress arousal (much like physical responses to lie detector tests). Subjects were also asked to express their feelings verbally while they were viewing the videos, which subsequently led Ulrich et al., (1991) to conclude that the participants' rate of stress recovery was faster for those who had viewed the video of natural settings in contrast to the urban settings.

The consistency of the research findings that suggest faster healing responses—as well as a more positive mental attitude in the presence of plants (or even the photographic or video representation of plants)—has been attributed by Korpela and colleagues to an *emotional affinity* toward Nature and natural scenes. In 2002, Korpela, Klementtila, and Hietanen discovered that subjects reacted more quickly and accurately to "anger" stimuli when primed with urban scenes but when primed with natural scenes, subjects reacted more quickly to "joy" stimuli. This caused researchers to conclude that exposure to Nature generates positive affect, while exposure to urban scenes arouses negative affect. In 2004, Hietanen and Korpela substituted faces for words, and the natural scenes varied as to the level of their restorative power and preference, but again positive emotion was associated with restorative nature and negative emotion was associated with non-restorative nature. Positive emotional benefits were discovered by the presence of roadside vegetation with respect to the levels of automobile driver anger and frustration by Cackowski and Nasar (2003), and in 1992, Shoemaker, Reif, and Bryant investigated the more personal role that flowers played in the bereavement process. This suggested that the presence of plants acts as a mitigating factor in situations involving escalating emotional distress.

Perhaps one of the most critical aspects of space exploration is the ability of crew members to be able to perform assigned tasks effectively, and the presence of plants has been shown to boost those abilities. In 1996, Lohr, Pearson-Mims, and Goodwin tested students in a standard college computer lab where plants were present or absent. The researchers discovered that their subjects were less stressed (as measured by taking systolic blood pressure readings), and performed better on a computer-oriented reaction-time test. They also reported higher levels of focus and attentiveness in the conditions that included the plants. Eight years later, Shibata and Suzuki (2004) found a similar result, when they assessed subjects' task performance by using a word-association task and their mood in three conditions as follows: a room with plants, a room without plants, and a room with a magazine rack (control). The reseachers reported that female participant task performance was improved when a plant was placed in the room, as compared to other room conditions. The researchers also suggested that this improvement in mood and task performance could have been affected by the level of compatibility between the task and the environment as well as the posibility that females demonstrated a high reactivity to the affective source.

Since there is evidence that *viewing* plants (or being in the *presence* of plants) has been found to improve individuals' mood, attention, focus, feelings of joy and happiness, etc., what other benefits might humans derive from *interacting* with plants (or Nature in general)? In 1976, a study by Langer and Rodin was conducted in nursing homes for the elderly where the residents were given one of two levels of control, with low control being designated as plant care in the common areas being conducted by the nursing home staff or high control being designated as plant care being conducted by the residents themselves. When behavioral and self-report measures regarding the residents were assessed, the researchers found that levels of alertness, active participation, and general well-being were all positively influenced by the high control condition. A follow-up study indicated that these benefits appeared to be maintained over time, as indicated by nurses' ratings as well as medical and mortality data as well (Rodin & Langer, 1977).

Such social benefits of having and tending plants may be generalizable to crew members who are deployed on long-term Space missions—particularly as they relate to the establishment and maintenance of community gardens. Community gardens have been identified as a pro-social factor within a community (Waliczek, Mattson, & Zajicek, 1996), within schools, urban settings, and long-term detention settings (Sandel, 2004); as well as on interracial relations (Shinew, Glover, & Parry, 2004). Further research has indicated that the social benefits of interacting with plants were observed to increase social interaction when plants were introduced into cafeteria settings (Rice, Talbott, & Stern, 1980). Research also suggests that plants mitigate violence and aggression among residents when introduced via "green spaces" into inner cities. Plant-placement in these unstable locales was observed to reduce levels of fear, and the exchange of fewer incivilities (Kuo & Sullivan, 2001).

CONCLUSION

Within and beyond the Earth's biosphere, plants have been typically regarded as a "natural resource" and a "source of nutrition" for humans and non-humans alike; however multiple research projects have indicated that plants are far more than just sources of economic convenience or a fallback for basic human nutritional needs. Plants provide psychological benefits for individuals who are traveling and living in Space for long durations of time in isolation and extreme environmental conditions. On Salut 6, decorative plants were introduced into the mission to provide crew members with support (Salyut-6 Experiments, 2019). The first experiment to directly involve cosmonauts in the tending of salad crops for the consumption of the crew members was conducted on the Mir Orbital Outpost. Lada continues to provide plant cultivation and consumption on the International Space Station, but no direct assessment of the effectiveness of these interactions has been yet conducted. It is worth noting that both American and Russian crew members have commented that cultivating plants in Space differs from the performance of the usual mission operations, because it is interesting, fun, and provides them with a link back to Earth. According to Kalari and Zaletin, 2001, "Today is the day that for the first time man eats plants that he grew specifically for that purpose in space . . ."

Plants are alive and ever-changing in their growth, and they provide an opportunity for crew members to be able to control something in their surroundings. This seems a natural affordance with benefits both on Earth and out in Space (in contrast to unnatural affordances made of metal and plastic, and the mechanized environment constructed out of necessity for space flight). Green leaves can provide psychological comfort as well as relief from monotony and boredom; plants also *depend* upon crew members for their survival and ability to grow and flourish, suggesting that part of the psychological benefit is the occupation with caregiving chores required to raise living things, rather than tending non-living machines (which can offer no biological or sentient feedback). In 1995, a simulation of a long duration space mission included the cultivation of a crop of wheat using the Svet greenhouse to test the Mir 22 Greenhouse II Experiment's protocol of harvesting two-thirds of the plants. During the post-experiment debriefing, Alexander Andruskov, who was the senior crew member, commented that the individual wheat plants had come to be considered as additional *crew members* on the mission, so that harvesting them was akin to "killing a team member" (Ivanova et al., 1988; Yampolskii, 1996). What a startling admission! This reaction may have helped to explain the crew's response during the experiment on Mir, when they only collected leaf sections—rather than harvesting the entire plant as they had been trained to do, which skewed the experimental results (Levinskikh, et al., 2000).

Yet many questions remain unanswered: What types of plants are best suited to space cultivation to meet both non-nutritive and nutritive goals? How will crew members come to regard space plants? As only a food source? As a pet? Elevated to the status of a fellow crew member? What degree of automation of plant care (if any) should be designed? How much creativity and initiative should be designated to the crew members? (Bates et al., 2010).

And one parting thought . . . if or when the crew members perceive plants as fellow crew members, how might the mortality rate of non-human crew members such as plants impact the psychological well-being of the crew? Thinking even further "outside the box", will the underlying *cause* of the death of the plants (either by some technological failure or by human error or oversight) impact the psychological well-being of the crew? If it does, to what degree? Like replacing your child's pet (secretly) when it dies (to prevent psychological distress due to emotional attachment and separation anxiety), would openly (or secretly) replacing plants viewed as essential companions mitigate or complicate any potential psychological distress being experienced by the crew members? If so, to what degree? (Markey & Meinecke, 2019).

If there is an innate relationship that exists among all of the species that share the biosphere on Earth, and if humans experience grief when a dissimilar species perishes forever here on Earth, will humans exponentiate that grief by accidental or unethical expansion beyond the planet's biosphere? These are the challenges of space travel facing a species in conflict; humans cannot imagine not exploring the cosmos . . . but too, humans cannot bear the possibility that, having so fervently touched the stars, humans brought with them a disease the cosmos could not survive.

REFERENCES

Bates, S., Gushin, V., Bingham, G., Vinokhodova, A., Marquit, J., & Sychev, V. (2010). Plants as countermeasures: A review of the literature and application to habitation systems for humans living in isolated or extreme environments. *Habitation (Elmsford, N.Y.)*, *12*(1), 33–40. doi:10.3727/154296610X12686999887201

Borenstein, J. (2017). Research involving human subjects. *CITI Program*. Retrieved from https://www.citiprogram.org/

Cackowski, J. M., & Nasar, J. I. (2003). The restorative effects of roadside vegetation.Implications for automobile driver anger and frustration. *Environment and Behavior*, *35*(6), 736–751. doi:10.1177/0013916503256267

Cleland, C. E., & Chyba, C. F. (2002). Origins of life and evolution of the biosphere. *Astrobiology Magazine*, *32*(4), 387–393.

Fjed. (2006). *Do plants in offices promote health?* Retrieved February 2006, from http://www.plants-for-people.org/eng/science/office_health.doc

Gazenko, L. G., Myasnikov, V. I., & Uskov, F. N. (1976). Behavioral control as a tool in evaluating of the functional state of cosmonauts in flight. *Aviation, Space, and Environmental Medicine*, *47*, 1226–1227.

Glenn Research Center. (2015). *Newton's laws of motion*. Retrieved on March 6, 2019, from https://www.grc.nasa.gov/www/k-12/airplane/newton.html

Grigoriev, A. I., Kozerenko, O. P., Myasnikov, V. I., & Egorov, A. D. (1988). Ethical problems of interaction between ground-based personnel and orbital station crewmembers. *Acta Astronautica*, *17*(2), 213–215. doi:10.1016/0094-5765(88)90026-4

Heerwagen, J. H., & Orians, G. (1986). Adaptations to windowlessness: A study of the use of visual décor in windowed and windowless offices. *Environment and Behavior*, *18*(5), 623–639. doi:10.1177/0013916586185003

Hietanen, J. K., & Korpela, K. M. (2004). Do both negative and positive environmental scenes elicit rapid affective processing? *Environment and Behavior, 36*(4), 558–577. doi:10.1177/0013916503261391

Ivanova, T. N., Kostov, P. T., Sapunova, S. M., Dandolov, I. W., Salisbury, F. B., Bingham, G. E., . . . Jahns, G. (1988). Six-month space greenhouse experiments – a step to creation of future biological life support systems. *Acta Astronautica, 42*(1-8), 11-23.

Kalari, A., & Zaletin, S. (2001). *Mir space station: The final mission* [video]. Video of the "Greenhouse 6" harvest.

Kanas, N. (1998). Psychiatric issues affection space missions. *Aviation, Space, and Environmental Medicine, 69*(12), 1211–1216.

Kanas, N., & Manzey, D. (2003). *Space psychology and psychiatry*. Dordecht, The Netherlands: Kluwer Academic Publishers.

Kanas, N., Salnitskiy, V., Weiss, D. S., Grund, E. M., Gushin, V., Kozerenko, O., ... Marmar, C. R. (2001). Crewmember and ground personnel interactions over time during Shuttle/Mir space missions. *Aviation, Space, and Environmental Medicine, 72*(5), 453–461.

Kaplan, S. (1987). Aesthetics, affect, and cognition: Environmental preference from an evolutionary perspective. *Environment and Behavior, 19*(1), 169–182. doi:10.1177/0013916587191001

Korpela, K. M., Klementtila, T., & Hietanen, J. K. (2002). Evidence for rapid affective evaluation of environmental scenes. *Environment and Behavior, 34*(5), 634–650. doi:10.1177/0013916502034005004

Kozerenko, O. P., Gushin, V. I., & Sled, A. D. (1999). Some problems of group interaction in prolonged space flights. *Journal of Human Performance in Extreme Environments, 34*(5), 123–127.

Kuo, F. E., & Sullivan, W. C. (2001). Aggression and violence in the inner city: Effects of environment via mental fatigue. *Environment and Behavior, 33*(4), 543–571. doi:10.1177/00139160121973124

Langer, E. J., & Rodin, J. (1976). The effects of choice and enhanced personal responsibility for the aged: A field experiment in an institutional setting. *Journal of Personality and Social Psychology, 34*(2), 191–198. doi:10.1037/0022-3514.34.2.191

Levinskikh, M. A., Sytchev, V. N., Derendyaeva, T. A., Signalova, O. B., Salisbury, F. B., Campbell, W. F., ... Jahns, G. (2000). Analysis of the spaceflight effects on growth and development of super dwarf wheat grown in the svet greenhouse. *Journal of Plant Physiology, 156*(4), 522–529. doi:10.1016/S0176-1617(00)80168-6

Lohr, V. I., Pearson-Mims, C. H., & Goodwin, G. K. (1996). Interior plants may improve worker productivity and reduce stress in a windowless environment. *Journal of Environmental Horticulture, 14*(2), 97–100.

Markey, M. A. (2019). *The Theory of Astrophilia.* Submitted for copyright to the United States Copyright Office on January 1, 2019.

O'Grady, M. A. (2016). *The theory of biophilia* (Reg. No. TXu 2-032-922). United States Copyright Office. Retrieved from https://www.copyright.gov/

Quinn, S. (2009). Plants as important in space as on earth. *Voice of America News.* Retrieved on March 7, 2019, from https://www.voanews.com/archive/plants-important-space-earth

Rice, C., Talbott, J., & Stern, D. (1980). Effects of environmental agents on social behavior of patients in a hospital dining room. *Hospital & Community Psychiatry, 37*(2), 128–130. doi:10.1176/ps.31.2.128

Rodin, J., & Langer, E. J. (1977). Long-term effects of environmental agents on social behavior of patients in a hospital dining room. *Hospital & Community Psychiatry, 37*(12), 897–902.

Sallin, R. (Producer), & Meyer, N. (Director). (1982). *Star Trek II: The wrath of Khan* [Motion picture]. USA: Paramount Studios.

Salyut-6 Experiments. (2019). *Global Security.org*. Retrieved on March, 9, 2019, from https://www.globalsecurity.org/space/world/russia/salyut6_experiments_seriespiloted.htm

Sandel, M. H. (2004). Therapeutic gardening in a long-term detention setting. *Journal for Juvenile Justice Services, 19*(1 & 2), 123–131.

Semantics. (2019). In *Dictionary.com*. Retrieved on March 7, 2019, from https://www.dictionary.com/semantics

Shibata, S., & Suzuki, N. (2004). Effects of an indoor plant on creative task performance and mood. *Scandinavian Journal of Psychology, 45*(5), 373–381. doi:10.1111/j.1467-9450.2004.00419.x

Shinew, K. J., Glover, T. D., & Parry, D. C. (2004). Leisure spaces as potential sites for interracial interaction: Community gardens in urban areas. *Journal of Leisure Research, 36*(3), 336–355. doi:10.1080/00222216.2004.11950027

Shoemaker, C., Reif, D., & Bryant, C. D. (1992). The role of flowers in the bereavement process. In D. Reif (Ed.), *The Role of Horticulture in Human Well Being and Social Development: A National Symposium* (pp. 43-46). Portland, OR: Timber Press.

Stemwedel, J. D. (2015). The philosophy of Star Trek: Is the prime directive ethical? *Science.* Retrieved on March 6, 2019, from https://www.forbes.com/sites/janetstemwede/2015/08/20/the-philosophy-of-star-trek-is-the-prime-directive-ethical/#2c8dbd382177

Ulrich, R. S. (1979). Visual landscapes and psychological well-being. *Landscape Research, 4*(1), 17–23. doi:10.1080/01426397908705892

Ulrich, R. S. (1984). View through a window may influence recovery from surgery. *Science, 224*(4647), 420–421. doi:10.1126cience.6143402

Ulrich, R. S. (1986). Human responses to vegetation and landscapes. *Landscape and Urban Planning, 13*, 29–44. doi:10.1016/0169-2046(86)90005-8

Ulrich, R. S., Simons, R. F., Losito, B. D., Fiorite, E., Miles, M. A., & Zelson, M. (1991). Stress recovery during exposure to natural and urban environments. *Journal of Environmental Psychology, 11*(3), 201–230. doi:10.1016/S0272-4944(05)80184-7

Waliczek, T. M., Mattson, R. H., & Zajicek, J. M. (1996). Psychological benefits of community gardening. *Journal of Environmental Horticulture, 14*(4), 204–209.

Watanabe, S. (Ed.). (2007). Life's working definition: Does it work? *NASA's Astrobiology Magazine.* Retrieved on March 6, 2019, from https://www.nasa.gov/vision/universe/starsgalaxies/life%27s_working_definition.html

Wise, J. A., & Rosenberg, E. (1988). *The effects of interior treatments on performance stress in three types of mental tasks.* CIFR Technical Report No. 002-02-1988. Grand Valley State University, Grand Rapids, MI.

Yampolskii, L. (1996, January). Three on the barrel, not including the wheat. *Ogonyok, 5*(4436), 4-5. (in Russian)

Chapter 3
How Taxonomy Steals Reverence

ABSTRACT

Chapter 3 is about classification and the paradox of losing one's positive regard for a thing by being able to describe it well. The philosophy behind the need to recognize what matters constantly versus the curiosity to discover what matters even once has a long and anxious history in many cultures, several predating Western thought. This chapter posits that the human habit of looking at many unique individuals as more similar than different allows them to remember what they cannot bear to forget—at the expense of ignoring the essential differences they will never remember.

INTRODUCTION

"Comparison is the thief of joy" – Theodore Roosevelt

What do the authors mean by *taxonomy*? What does taxonomy have in common with reverence? Reverence is the inability to classify a feeling of wonder. Taxonomy, on the other hand, is the "orderly classification of plants and animals according to their presumed natural relationships" (Taxonomy, 2020). The scientific principles arising from the classification system guide theory and practice. Often, it seems, society would rather preserve its classification system than its planet. Wonder for Nature (and the speechlessness that goes along with that), does not seem in fashion anymore; documenting what is

DOI: 10.4018/978-1-7998-4408-2.ch003

about to go extinct has become more popular. Mankind has no intention of changing his ways.

What stands out, however, is how similar this attitude is to a famous line from *A Christmas Carol*:

If these shadows remain unaltered by the Future, none other of my race," returned the Ghost, "will find him here. What then? If he be like to die, he had better do it, and decrease the surplus population. (Dickens, 1995, pp. 81-82)

This profound line from an endearing holiday story, summarizes the almost teleological[1] attitudes that human sciences hold toward Nature. Every living being must fit a scientific category, and any being that does not fit one is either feral or ill. This attitude toward living things is tragic enough in the physical sense; in the mental sciences, it is clearly banal in its implications (Hergenhahn & Henley, 2014; Pawlett, 2007). The *will to be* is not constrained by the need to fit a classification system, in much the same way as "the quality of mercy is not strained" by the need for a universal mercy (Shakespeare, 2010, 4.1.189). The individual is burdened with the *will to be* not the fear of fit, such that, if there were a science of wonder, the individuals which could not be fit to any category would dominate the field and its data.

This chapter is about the practice of taxonomy (the classification of just about anything into pigeon holes wherein unique individuals must pretend not to be unique if they hope to find fit in society). This chapter proposes that the ideal of individuality and the categorization of individuals is the myopic pursuit of antithetical goals. This same paradoxical pursuit of opposing goals underpins the current disregard for the living environment's survival, and a similar disregard for the depression and helplessness of human children.

ARE YOU A GOOD ILLNESS OR A BAD ILLNESS?

What is *disease*? What is *health*? The authors ask this because health to one species is quite often a disease to another species. And some diseases are fortuitous, because, since they do not extirpate this financially focused species completely, they can serve as inarguable reasons to wage an economically beneficial war against them. The comparison of similars permits the explosive growth of categorical differences. Similitude is about finiteness; categories are simply lots and lots of finite differences shared by an overarching similarity. Comparison is hurtful, in other words.

Theodore Roosevelt once said that constantly comparing one thing to another thing (to see which thing makes one more happy) was the reason for being unhappy (Cooper, Gustafson, & Salah, 2013). The mental health community says that the struggle to keep track of which things are healthy and which things are unhealthy is itself unhealthy (Brauser, 2017). The famous Viktor Frankl called this effect *paradoxical intention* (Frankl, 2006). Taxonomy is really about that—the practice of classifying everything suggests humans need to know what everything is and everything does, othewise humans won't trust it.

THE WAY OF NATURE

The way of Men is easy to understand and predict (humans just described some of it). Humans keep track of everything that exists and sort it into safe and unsafe, useful and unuseful, sustainable or futile. If it looks like this and it does that, men can teach you all about it—why it's here, why it went extinct, and how to make a living off that particular thing before it goes extinct.

The way of Nature is not so simple. Western thought has been occupied for millennia by primarily Western principles, so Western conclusions have been predicated on a subset of possible explanations (Sundararajan, 2015). But in the East, the pursuit of wisdom is less about enduring vanity and certainty than it is about humility and uncertainty. In fact, entire empires once selected their officials and public servants using Eastern philosophical principles like Confucianism as their selection criteria (Meinecke, 2017). As an example, the Chinese philosophy called Taoism (or Daoism) is sometimes called "the way of nature" (Davis, 2018, para. 4). It stresses awareness without thinking. It favors an instinctual drive to work together for good, rather than a competition over scarce resources. It stresses harmony *with* Nature rather than subduing Nature. Zen Buddhism is similar; it tends away from generalizing, categorizing, measuring, and ranking everything (which is pretty much the goal of science), and emphasizes instead how essential each minute thing is to the whole (Nagatomo, 2020). There is a concept called *prajna* in the East. Although it is translated without much understanding in the West (as a kind of judgmental appreciation), it is defined more aptly as a kind of gentle appreciation or acquired insight in the East. Some call it "non-discriminating knowing" because it is a non-dualistic perspective of Life that accepts things in their "suchness" and "isness" (just as they are, an undifferentiated condition called *tathata*), rather than concealing some deeper meaning knowable only to sacred people and sacred texts (Bose, 1996; Ives, Abe, & Hick, 1992; Nagatomo, 2020).

Even the most enlightened authority (including Buddha) views his intellect as empty or nothing relative to the Truth (and in earnest humility too), which is a very good way to humble oneself before an unthinking Nature. And though, in Europe, the Middle East and the West, sacred writing is valued above human life itself (be it spiritual or secular verse), Hakuin relates that, for the seeker of enlightenment in the East, one's treasured writings are no more sacred than toilet paper compared to the object of that pursuit (Yampolsky, as cited in Bose, 1996). What inspires remains forever more sacred than what it inspired.

What is a disease? Is the way of Nature a disease? Or is the way of Man a disease that denies that Man should be classified as a disease? The authors submit that the Western regard of Nature is an effort to distinguish significance and purpose while confining itself primarily to its own reduced set of colonialist, anthropocentric values and distinctions. Is it hard to find God in Nature each day because humans are too certain God looks like one of them? The authors of this book have undertaken the task of conducting studies to see if this might be so. What if the idea of God were apparent in everything humans adore? It's hard to say. But one thing is certain: when the things you can manifestly ascertain seem to be in the way of things you consciously struggle to see, it is easier *not* to trust what is already manifest. That way, you can withhold your trust and your reverence for Life until your suspicious, invariant criteria think you've recognized something strangely familiar you can trust. There is even a famous effect called the uncanny valley, which arises when you are about to trust something that looks too human. Freud's term for the uncanny was *unheimliche*—unhomelike—because it seems frightening (even foe-like or disease-like) yet strangely familiar (Freud, 1919; Freud, 1920; Yamamoto, Tanaka, Kobayashi, Kozima, & Hashiya, 2009).

MUCH ADO ABOUT TAXA

This section will begin with some definitions. What is taxonomy? According to the Oxford English Dictionary, it's the science of classification—and especially the classification of living things (Taxonomy, 2019). But according to E. O. Wilson (2016), someone widely acknowledged as an expert in taxonomy, it's become a way to further various agendas, such as that of the *new conservation*. What is that, you ask? It's the new push for industry-focused or politically focused conservation efforts, in which Nature apparently exists for the welfare of the human species—and that's why humans should take measures to preserve the parts of Nature useful to Mankind. The problem with this mindset is that the whole purpose of Life and of staying alive becomes skewed—such that the goal of the preservation of Nature is to prevent the extinction of civilization—not to prevent the extinction of Nature. Perhaps more troubling, is that the physical kind (the biological species known as Mankind) is no longer what humans are protecting from going extinct (because primitive people everywhere are going extinct, but not *civilized* people who are actually thriving). Humans have even dubbed the modern era, the *Posthuman* era, partly because the struggle to produce information has become more precious than the struggle to produce humanity. Curiously, the new focus is not to prevent the unrecoverable loss of the human species, but to prevent the unrecoverable loss of human thoughts and works (since the *mass production* of human ideas and products seems to be the human idea of "Nature"). Humans seem content to transfer their engrams into lifeless computers, so humans don't have to die when their bodies die, like every other species does. This lack of mercy for living things is in the authors' research—it's called *Justice for Nature*. In that research, the survival of "jobs" has become a higher priority for humans than the survival of the Earth's air, water, soil, plants, or animals. And rather than feeling contrition when asked to give a little back, many folks seem quite angry that humans cannot have what little remains too. This theme is also part of E.O. Wilson's work on how to save the planet from imminent extinction.

How did humans arrive at this strange mindset? How could humans compare "a job" to Life itself? Isn't that like comparing wax apples to real oranges? More concerning, however, is the implication of such a bizarre shift away from the fundamental physiological urgencies of a biological species, toward "ideological urgencies." What would make humans defend an idea at the expense of a living being? This is some very scary thinking (the survival of jobs and institutions before the survival of Life on the planet). It is an odd thing that what humans call a *job* merits more of their reverence today than the living creatures being displaced, exiled and extirpated by their jobs. Not so very long ago, humans were in awe of the natural world, even worshipped it, because humans were so grateful it was here. Now—it's in their way.

So, now can the reader see how taxonomy might be stealing human reverence? Look at this shift in human thought from life-focused to information-focused, from survival-focused to task-focused (Zeigarnik, 1927). A job is not a living thing; a business is not a living thing; in fact, no abstract institution created by the human species has ever been alive (let alone are any of their thoughts or works in danger of extinction; Isaiah 55:8). None of these things are species. That is why humans invent concepts such as "corporations"... to *embody* the idea of a living thing (but eerily using an abstract concept—because human concepts are immune from the ravages of biology). And whatever is immune from the ravages of biology, somehow replaces whatever reverence for Life once reigned over all of that biology (Meinecke, 2018b, 2018c). Nonetheless, even if these human things cannot perish, humans cherish them as much as (or more than) things that do (including each other, ofttimes). Humans do revere *categories* of things. some living and some not (and perhaps the whole idea is to revere things that are not alive, which humans once called *idolatry*). For example, humans revere "their species" over other species, and "their stuff" over anything alive, it seems. So, even if humans do not revere Life itself in its endless variety of expression, humans do seem to revere "collections of things" of categorical value—some of which happen to be alive, but most of which are not alive.

In this short section, the authors propose the idea that "taxa" (categories) rob individual beings of their value, much like *thieves of joy* as Roosevelt said it so eloquently. This happens reliably, whether that thing humans steal value from happens to be a human being or something else. Why is that? What do the authors mean by a *theft of value*? Very simply, nothing of infinite value would be put into a category, would it? Why bother to create a category of just one thing? No; each living thing is one-of-a-kind, a pearl of great price (Matt. 13:46). And a "category" is not a thing at all. It is just an abstract idea that you can conveniently put real things into—much like a barrel. Even though an apple barrel isn't an apple, it can hold lots of apples, and after a time it is enough to see the barrel and imagine humans have seen the apples. A nice benefit is immortality. Because categories are not real, they don't have to perish like real things do. So, in time, because humans usually see the apple barrel full of invisible apples (rather than what's inside), humans eventually associate the "apple barrel" with lots of tasty apples. Humans may even salivate at the sight of a wooden barrel. Even worse, the next time humans are hungry, and humans are looking for "apples", humans might look for "barrels" instead, and forget what an apple looks or tastes like. Trust the authors on this as research psychologists; that is precisely how mental association works. And sadly, in the process, the value humans understood to be in the image of an apple is transferred (incidentally) to the image of a barrel (Pavlov's dogs, 2019). Does this sound a bit like Genesis here? In fact, it is probably impossible to form an association without losing the distinct value of each of the things you think are associated.

Now it's time to get back to the main topic and the title of this chapter. In the science of taxonomy, a *taxon* (plural *taxa*) is just an empty "barrel" into which you can conveniently put just about anything. Humans use barrels in programming too, but humans call them "variables" and "pointers" instead of barrels or categories or species. But they aren't real things; in fact, they may not even contain anything meaningful, or actually point anywhere at all (just think of "dead hyperlinks" and you'll get the idea). But "variables" are very convenient to represent averages of things, and to indirectly point just about anywhere (even if they often point to nothing at all). One of the authors can speak from education and experience on the subject of programming—and how this "reverence" for information processing can make you miss out on the living persons in your life that you ought to have revered. He did, and grief was the result.

Another thing that taxonomic categories do, in addition to stealing the *value* of the things humans stick into categories, is to modify the *meaning* of the things humans stick into those categories. No two things are exactly alike in Nature, let alone do two (or more) things fit nicely into even one category (per Darwin and many others; Darwin, 1876; Unterseher, Westphal, Amelang, & Jansen, 2012). There is quite a lot of evidence of this (the problem of quantum nonlocality, if nothing else; Nonlocality and entanglement, 2019). So, there is a lot of solid argument against putting unique things into heaps so you can refer to them as being in a category. But if humans put more than one thing into a category anyway (and humans can), it does become a lot easier to refer to the overarching category (or group trait or containing species) than to each special *individua*l that happens to be aggregated into that handy heap. In fact, if humans put even one creature into a category, the *creature's* unique qualities will be lost—in favor of the *category's* qualities (Biernat & Danaher, 2013). The result is considerable existential conflict (and see quantum nonlocality once again, for the microscopic equivalent). Living things apparently do not like to be reduced to a bunch of identical things, unlike their lifeless human commodities (which humans go out of their way to make identical). Living things like *to be perceived*, as Bishop Berkeley put it (Downing, 2013). As you can see, one of this author's favorite quotes is attributed to President Theodore Roosevelt: "Comparison is the thief of joy" Why did the author choose his insight for this chapter? Well, if ever humans see two things as similar, humans compare them to see which one is "better" or "more representative" of the similarity (because there can only be one, really). When humans do this, it steals the worth of each individual thing in order to try and establish the worth of the category. It steals joy. Perhaps the authors should summarize this as a sort of principle:

A concept is a thief of whatever one puts inside of it

Whether humans put 10 apples into the same category, or even one apple into that category, this seems to be the case—the idea of an apple is overshadowed by this general idea. Ever after, humans protect the survival of their category, not what it contains.

On this same topic, Dr. Philip Zimbardo is famous for using barrels as an example to argue against the genetic basis of evil. He asks, are some people just plain *born* bad apples? Or, he posited in *The Lucifer Effect*, do average apples go bad because humans put them in barrels which *guarantees* good apples will go bad? (Zimbardo, 2007). He is not alone either; he is in good company with prominent thinkers, such as Arendt and Adorno and Milgram—and many others. But using "their" taxonomy, some species are "created evil" and some species are "created good"; not surprisingly, humans put "their species" into the good category and every other thing in Nature into the evil category. Why is that? Just maybe, neither humans (nor anything else) are evil by nature. Just maybe it is *the practice* of putting things into categories that makes some of them seem better than others. After all, if you put several pearls of great price on the same shelf at the same price, won't you wonder if one is worth more? How can you tell which one to purchase if you must go and sell everything you have to buy just one of them? Whenever humans create the idea of evil, whenever humans put things into that category, humans try and figure out which is "more evil" than the others. That's just weird. Humans rank everything from best to worst, greatest to least, god-like to demonic, whenever there are two or more of the same thing (Figure 1). Spurgeon (2013) once wrote, "Of two evils choose neither. Don't choose the least, but let all evils alone" (p. 341). Maybe the belief that there are categories for good and evil, makes anything humans accidentally put into them seem good or evil. Maybe they are just things, after all.

Figure 1. The Great Chain of Being (Public domain)

THE SORITES PARADOX

Next the authors should explain a very old concept, because this very old concept is very pertinent to this chapter. The ancient Greeks had their own way of looking at the problem of taxonomy. They were so puzzled by what constitutes a group of something, they called it the Sorites Paradox—from the Greek word *soros* or heap (Hyde, 2011). The basic approach was to ask how many grains of salt it takes to make a heap of salt (or how many grains you can take away and still have a heap of salt). Does one grain of salt make a heap? Do two grains of salt make a heap? How many grains does it take?

This research suggests that a heap begins... not when a certain number of grains is reached... but when it becomes *easier* to see “a heap” of that thing than each grain of that thing (Vrobel, 2011).

The Sorites Paradox, curiously enough, persists even in this present era. But it is less about salt and heaps of salt, than it is about living individuals and heaps of individuals these days. Grains of animals are called *members*, and heaps of animals are called *species*. Grains of people are called *individuals*, and heaps of people are called *nations or races*. The new questions ask how many people it takes to make a people, or how many different ways people can behave before humans admit they behave like nobody else behaves. The new heaps are made of psychological granules, though, instead of being composed of real people, real salt, or real sand. Psychological categories heap unique individuals into categories, just as humans have always heaped individual “cattle” and “sheep” into categories. The individuality, the hopes, and the dreams of the living individuals become harder to see, so that the categories can become easier to see and treat using various procedures (NHS Choices, 2013). The same logic applies: if all humans are special, why do humans need more than one of each? How many people of each kind are needed before each kind is perceived as precious and irreplaceable? Isn’t one sufficient? Humans also approach this in reverse, just as the ancient Greeks approached it—how many members of a species can humans afford to lose but not lose their species? One surviving male and one surviving female? What if one of them isn’t fertile? What then?

The answer to how many of a kind can humans lose before humans lose that kind should be: None. Each member of a species is its own unique species, and no other member is exactly like it. Tragically, the fear of losing living individuals only becomes critical when the idea of their *species* seems to be in danger of perishing—not when each irreplaceable individual is about to perish forever (Ceballos et al., 2015). Even the fear of Death itself has become overshadowed by the global competition to see which civilization can make everyone else perish before they do (Freud, 2019). Human society worries mainly about the death of its ideas, not how many people die in the process of revering those ideas (Meinecke, 2018b, 2018c).

THE MYTH OF SPECIES AND MEMBERS OF SPECIES

Charles Darwin (1876) may have been among the first to observe that there is no end to the species. He was certainly famous for pointing it out—he made it plain that he did not see any clear demarcation between one variety and another. Yet, to this day, naturalists continue to echo his awe. For instance, new researchers name their articles with words describing their surprise: *3,000 Species and No End – Species Richness and Community Pattern of Woodland Macrofungi in Mecklenburg-Western Pomerania, Germany* (Unterseher et al., 2012). In Darwin's time, there was a great change in human thinking. There was a startling realization regarding just how majestic Nature really was (it seemed to be symbolic of an endless struggle to variety, not categories of similar immutable things). Even within what appeared to be just one variety of one species, each leaf on the same branch of the same tree exhibited individual differences, in a struggle to survive! Darwin (1876) said it this way:

Certainly no clear line of demarcation has as yet been drawn between species and sub-species... These differences blend into each other by an insensible series; and a series impresses the mind with the idea of an actual passage... from these remarks it will be seen that I look at the term species as one arbitrarily given, for the sake of convenience, to a set of individuals closely resembling each other... the term variety... is also applied arbitrarily, for convenience sake (p. 31).

Curiously, even though the "father" of the idea of speciation and evolution remarked on the arbitrariness of taxonomical obsession, humans try harder today to fit into a "known category" than to carve out their own category. This be will looked at again when the topic turns to containers and contents, organisms and organs, and pilots and vessels; but for now, consider the obvious irony of trying to assert that more than one of a "unique thing" is scientifically possible.

Perhaps it's time to play a language game—in Wittgenstein fashion—because a mind game is a great way to come to the end of a fallacious conclusion before you accidentally believe in it (Wittgenstein, 1965). When is an individual a species? Or, how many unique things does it take to make a unique thing? The answer is not hard; the answer is one. The answer, though, requires not thinking about anything living as a "member" of anything (living or non-living). If logic cannot do it with salt, why do humans think logic can do it with living things? The paradox seems to be less about minerals or animals, than to simply stem from the need to "refer" verbally to actual things (especially when they aren't here to refer to). That way, humans can point at something whether it is here or not. Once you see the reference, you do not need to see what it refers to. And when you can do that, you cease to need to see the continuing differences you can only witness if you never look away. Nature does not stay the same. Humans can never look away. But once you start looking at the name or trait or history of a man instead of the man, you cease to see a struggle for existence by a real man. Instead of reverence for the real creatures humans live with, humans let their concepts supplant the real creatures their concepts refer to (Meinecke, 2018a; Pervin, 1994).

It's time to backtrack. Maybe the fundamental problem is the notion that "categories" actually exist? Maybe that's what spoils anybody's hope of being irreplaceable? Well, when the literature is reviewed, that does seem the case. Biernat and Danaher (2013), like Sherif, Harvey, Hood, and Sherif (1961) argued that simple categorization, like group membership itself, is *sufficient to foster categorical disagreement, group hostilities, intergroup dehumanization, and intragroup jealousy* (Fiske, 2011). Simply saying that a category exists creates havoc, not order. Now living persons are more worried about fitting into an appropriate "category" than they are of being real people. Fitting more than one precious thing into just one heap of precious things seems to engender more categorical hostility than it does referential utility. The principle of being both separable and united is not sensible; however, to fit into a civilized culture, you must reconcile insensibility with common sense (Austen, 2019; Karakostas, 2012).

For instance, in westerns, when the living citizens give their lives to save the idea of a non-living town, the storyline and its concept of human sacrifice is held to be endearing (Rosemont, Sharpe, Bauman, & Hardy, 2000). Our natural compassion for living things that are at risk of perishing (such as people and animals), has been malformed by various concepts such as colonialism and expansionism. Humans now exhibit compassion for "ideas" (such as towns and animal species), as though humans are more afraid of losing their towns than of losing real human beings (Arndt, Routledge, Cox, & Goldenberg, 2005; Brosnan & Michael, 2014).

This can be looked at in reverse. How many individuals can one lose and still have an individual? The answer is easy—none. There is only one of each; there will never be another quite like each one. There is no such thing as a quorum of individuality. Those who have had children know this so intimately(losing just one child is like losing all of them), and those who have had beloved animal companions know this as well (you cannot simply replace a lifetime of friendship with another one). Because, when the living stop being precious and inscrutable, they become arbitrary grains of salt in a human heap. The trade-off to not viewing humans as one-of-a-kind, is that human industry can more conveniently work with heaps of people, instead of individuals—even if it means neglecting the very idea of individuation (Eliot & Scheffler, 2009).

So, back to the question: When is a living individual just a species of something else? The answer is, a living individual becomes a part of something "else" when that individual is no longer perceived as an individual; when any living thing is no longer insensible, inscrutable, unspeakable, it ceases to be the one and only member of its kind—and becomes a mental or verbal commodity that raises the collective value of an ideological group (Deleuze, Guattari, & Foucault, 2009). This chapter argues that biophilia is about realizing that nothing is a species of anything but itself; nothing is part of a frequency greater than one. In each moment, it is the only epoch of its kind, and there will never be another epoch like it. Even Bertrand Russell made this plain: natural language limits expression. It reduces the ability to describe the most infinitely special things to a known repertoire of vague propositions (Ludwig & Ray, 2002). In contrast, what cannot be uttered because it is beyond description, comes out not as intelligible words, but as awe, or tears, or trembling, or wonder, or grief—in any of an infinite variety of ways. What comes out is anything but a conventional word (because a word is a heap of whatever it is humans are struggling to name). As Heidegger might have put it, the thing-in-itself leaves one speechless (Spinelli, 2013). When humans try to honor the endless wilderness of the Earth's biosphere, all they can do is experience a feeling of Wonder; humans look at one another in bewilderment, astonishment, and gratitude (Muir & Highland, 2001).

HOW REVERENCE CAN RESTORE OUR AWE

If taxonomy steals reverence for things, maybe reverence can restore awe for things? Taxonomy is a skill, but reverence is like being overwhelmed. Classification is the capacity to identify, describe, and control the behavior of otherwise astonishing things. In the process, humans also control away their awe for them. Reverence, however, is a feeling of being overwhelmed by Nature, not being able to easily express how overwhelmed you are. That's hard to do if you think you know more about what you are in awe of, than it's trying to nonverbally tell you at the time. No, that's trying to expect awe. Awe is not something you can expect (and still experience the *complete* idea of awe anyway; Spinoza & Curley, 1994). When reverence is here, words fail as a verbal skill, and humans *gasp*. That's exactly how humans used to feel about Life in all its kinds, a thing humans used to call the *Breath of Life*. That's precisely how humans felt in the presence of truly natural Majesty. Muir is famous for reminding people to feel awe in the presence of an untrammeled Nature (Muir & Highland, 2001).

It's time to play make-believe. Pretend that humans are a group of little kids that have come upon a cliff at the edge of a hilltop campground (and humans aren't supposed to be this far from the camp). When humans stand as overawed youngsters on this hilltop, gaping over the ledge (and nearly trembling at the sight), what is it exactly that elicits awe? What makes a person feel the most reverent? Is it the fact that humans are looking down all of a sudden, on clouds humans are used to looking up at? Is it the angle? Or maybe it's because *every* angle makes a person dizzy (in a good way), that humans can only describe this moment as a kind of unnerving awe? The experience of awe for Nature doesn't happen in just one person who is really good at finding awe, nor is there a certain number required to find awe. No, awe for Nature (aka Biophilia) is like the approach of a spiritual entanglement, a happening that happens in every cell of what humans are, no matter how many humans there are (Keltner & Haidt, 2003). It's as though the entire body were part of honoring that moment, and the mind just a lucky observer.

So the principle of reverence ought to be that *nothing* should prevent one from feeling reverence, no matter how importunely God comes knocking at the door of one's regard (Luke 11:8). If humans wait until certain things knock before humans open the door to their regard, only certain things will bring them awe. The authors are even now in the process of constructing psychological studies using hidden object games, to study this phenomenon—of waiting for certain things before humans decide that something is worth waiting for. There is a lovely song about it, called *The Question* by Kevin Gould (K. Gould, personal communication, February 24, 2019). In that song, a little boy asks various grown-ups if they have seen God at all, but most are too busy to notice God. Only an old man gazing at Nature has seen God that day, and says, "Son, when I look around, I thinks I sees nothin' else" (Gould, 1976, stanza 6).

Unfortunately, the skill humans call taxonomy is actually the habit of constructing official categories of things humans will allow ourselves to believe in (before humans will believe in them). But what if nothing worth adoring will fit into a human category? What if the whole point of awe for Life is that *what* humans are most reverent toward is *one-of-a-kind*? What then? In that case, a category is the last thing you want to have. Categories are great for average things. But the thing about reverence, is that it begins where average things end.

If the skill called classification has stolen human reverence away, maybe an admission that nothing in which there is the ubiquitous breath of Life can be categorized or generalized (no matter the skill)—maybe that simple humility could help restore human reverence for Life? Maybe Darwin wasn't telling people not to believe in God after all. Maybe he had found a way to prove God exists... in the simple idea that maybe there is *no end to the ways* that God expresses Himself. And Maybe Darwin had uncovered the fact that a kind of infinite worth in infinite combination is all around—and within, too? So, if humans are this amazing, just imagine how amazing everything else on this awesome planet is? Maybe the complete idea of God is Awe! It is worth noting that Spinoza wrote an entire section on this very idea—the complete idea of God, even if it wasn't a very popular idea for his day (Damasio, 2003; Spinoza & Curley, 1994). Humans can only imagine *that idea*, unless humans let all of it completely in.

THE MYTH OF ASSOCIATION

Taxonomy steals a lot more than just human reverence, is the reader aware of that? Just as humans put "similar" things into categories (stealing the worth of each individual thing so humans can say there are "this many" of the same thing), so too humans say certain things *go with* certain things (even if they are remarkably dissimilar). These are really one idea, you know, frequency and association. Frequency reduces the worth of individual things using the idea of a *quantity* of the same thing, whereas association reduces the worth of individual things by suggesting that you need a *combination* of various things before you have one thing! (Meinecke, 2018b, 2018c).

For example, say humans need exactly one male and one female to make a couple (but it turns out that there are more females than males). The upshot is, some males will be forced to live alone for the rest of their lives, just because humans said you need one of each. Why do these human *moduli* exist, then, if they have no matching half? Simply because they do not *randomly* come in perfect pairs, many people will be incomplete, and the living pieces will endure decades of preventable loneliness. Therefore, an official association is more likely to *create* imbalance in a natural system, rather than ensure that everything out there has a perfect match (a "balanced" system). And humans note that among some "successful" species, an alpha male will even (selfishly) accumulate most of the fertile females, forcing all the other males to live out their lives frustrated and angry and alone. When that happens, in addition to lots of miserable males and equally miserable concubines, only one male among many will have progeny, reducing the gene pool to a less survivable subset of genes. (Kings and the well-to-do male among the human species used to do something very similar). From an evolutionary psychology mindset, however, humans try to see the good in this (even if there isn't any). Humans call that *survival of the fittest*. Humans reason that the most "fit" male will bear most of the offspring. That's great, but that also means the most *selfish* male will bear most of the offspring! Therefore, most of their *children* will be selfish too. Authors like Temple Grandin explain that humans are much better off with a random mutt who's bred when humans weren't looking, than an animal of good breeding whose consummations humans faithfully observed. And she provides ample evidence of how disastrous this can be (Grandin & Johnson, 2005). Authors like Dacher Keltner even suggest humans might actually have been a very good, non-hostile, even gentle species, and now humans are selfish and cruel because the genes humans continue to favor require that egotism and sexual frustration survive. Maybe humans didn't always practice what Darwin called the *unnatural selection* of everything, to create a *favored species* (Darwin, 1876; Keltner, 2009). What did Darwin mean by this idea of "unnatural" selection? He noticed that every other species seemed to practice natural selection (all except humans). He meant that trying to improve the species using the brain's idea of what ought to survive, is equivalent to an unnatural behavior rarely seen in Nature (and isn't it odd that the insightful Temple Grandin concurs?). And isn't it coincidental that the authors' research strongly suggests that humans seem far better off with *random selection* than even Darwin's observation of natural selection—especially if humans don't want to end up becoming a super fit AND super selfish species (and essentially that's what humans have become).

It's time to step away from the irony of associating certain living things with certain other living things to improve the lot of favored things. It's time to take a look at this imaginary associativity using *inanimate* things. Will humans see the same irony? As the author's father would have put it, "You betcha." For instance, if a woman has two pair of gloves (one brown and one black), and she loses one glove from each pair (and, by a stroke of blind luck, she loses one left glove and one right glove), she will *still* throw away *both* pairs—even though she has a perfect pair! Why? Because one is *brown* and the other is *black*. Surely she can't go out in public like "that" humans contend! Who says? Is there a glove fairy? Why would a woman with a perfectly good mind (and a perfectly good pair of gloves) think it a social *taboo* akin to incest, to wear mismatched gloves? Well, the gist of it all is that, once you go down that road ("this goes with that"), there's no end to what has to go with what. You see, the idea of a "matching" or kosher pair can also be:

- One right glove and one left glove constitute a pair
- Two gloves of the same color constitute a pair
- Two gloves made from the same material constitute a pair (e.g. cotton or leather)
- Two gloves with the same kind of stitching constitute a pair
- Two gloves with the same amount of wear constitute a pair
- And on and on…

This is prejudice, actually—not "association" (Biernat & Danaher, 2013; Meinecke, 2017). What do the authors mean, calling social appropriateness *prejudice*? Well, maybe this will help:

- A person with black skin doesn't "go with" a person with white skin
- A lonely female and another lonely female "do not go together" (God "meant them to be lonely")
- A person of "this" faith shouldn't associate with a person of THAT faith
- Human rights shouldn't apply to "animals" too (even if humans are clearly and scientifically animals, not something else)
- "Foreigners" shouldn't have the same rights as "us" (only things accidentally born on the same side of imaginary political lines "go together")

Do you see how ridiculous this can get? If humans assert that "this thing" goes with "that thing" (so that the only surviving combination adds up to "this favored thing"), humans are simply and prejudicially putting certain things together, while forcing other things to remain separate (unnaturally separate, one might add). Most species don't exhibit this bizarre behavior (and humans excel at it). Here is a short list of civilized mental biases from one of the authors' dissertations:

Can you spot the similarity between a pair of gloves and human bias? Maybe this will make it easier to see:

You see, be it living or be it clothing, humans think a pair of anything must "match" somehow. So, if humans had just one leg, humans would not laugh at people for wearing different things on different legs. It's because

Table 1. Types of Human Assessment Bias

Bias	Assumption
Inheritance	Genetic superiority
Race	Favored races
Ethnicity	Favored groups
Gender	Inferiority, social role
Age	Physical/mental inferiority
Intelligence	Mental superiority
Disability	Social liability

Source: Neglected by assessment: Industry versus inferiority in the competition for scarce kidneys, Meinecke, 2017

humans treat their two *different* legs as "a pair" that *both* left and right legs must wear the *same* color socks. Why? Who says? Can you see the irony? There is only one left leg. There is only one right leg. There is no such thing as "legs"—that is an idea which allows humans to group fairly similar things under one umbrella concept. It is simply a mental combination, a sort of prejudicial alchemy of Life, like saying things must or must not be a certain way for no feasible reason at all.

So, even if the woman has two *perfectly good* gloves, and even if one fits the left hand well and other fits the right hand well, and *even if her hands are shivering*, she will *not* wear them (unless they are of the same color and the same material and the same pattern of stitching, that is). Shivering is a physiological signal. Propriety is a mental demand. This thinking is a lot like how humans favor certain genders and certain levels of fitness (Meinecke,

Table 2. Types of Glove Assessment Bias

Bias	Assumption
Left/Right	Symmetry (at least this has a physical dimension)
Color	"Etiquette" (Why must the colors match? Why do humans scorn those who don't comply?)
Texture	*Hyperspecificity* in Grandin's wording (you have to get up really close to tell the difference)
Gender	Treating clothing as though there were female and male clothes, and enforcing a taboo against wearing the wrong "gender" of clothing
Age	Inanimate ageism (wear and tear treated like the young and old getting married)
Cost	The superiority of inanimate patterns (aka the fitness of the buyer)
Stitching	Similar to excluding others because of "abnormalities"

2017). Even if humans have two *perfectly* good people, and both are lonely, they must remain apart unless they "match." Who says? Don't loneliness and companionship match? And even when a man and woman "go together", they must also be:

- The same age
- The same race
- The same religion
- Have similar "interests"

Oh, and they'd better not be "related" or humans go bonkers (Freud, 1983). Curiously, it remains a troubling theological question, how exactly Adam and Eve's children avoided incest. Yet humans get really upset if two genetically related people are put together. Even more strangely, humans don't get upset when desperate people are excluded from even the leftovers of what humans enjoy, simply because they aren't "like us" (in the favored taxonomical group). Humans treat people a lot like gloves.

Nature, fortunately, is a little less picky when it's shivering outside in the cold (whether the living gloves it needs to stay warm "go with each other" or not). Nature will don the inappropriate to survive, no matter what people think about their habits, whereas people wear clothing because humans think humans would be "naked" without them (Genesis 3:11). If even God does not condemn us, who is he that condemns? Humans, though, go ahead and let their flesh shiver because "these things don't go together", or because humans maintain that "animals don't experience cold or pain like *humans* do"—even when an apparently unwed woman who is greatly with child comes to the door on a cold night, begging for *posada.*

CONCLUSION

Why don't humans honor Nature as much as humans honor their taxonomy of Nature? And even when humans do honor living things, why are humans so picky? In this chapter, the authors can propose a simple principle for Biophilia (reverence for Life):

Life is honored when all living things are honored, not by honoring certain things which happen to meet certain human criteria.

If humans don't have to memorize what to honor and what to disregard, everything that lives and breathes will find honor. No human should have to to worry whether he or she is alive enough to fit somebody's idea of Life. Neither should Nature. But too often humans have to be the "right" species to find honor. Too often humans have to be the best at something to find honor. No wonder both children and the elderly feel unworthy of being alive!

Posit, for a minute, that the idea of Life is simply any inspiring stimulus—not a certain pattern of stimuli that humans happen to ascribe inspiration to. With this simple idea, *any* natural stimulus can bring awe. Even an unexpected gust of wind lifting the hem of one's garment, can make one ask who touched you, and in that question bring one and all warm tingles up and down the spine. Isn't that a form of awe?

In this principle of reverence for Life, or awe of Life if you will, what *humans understand least* (not what humans are certain of) is made semantically manifest by an uncontrollable trembling in all that humans are. You can't wait for something specific, and find yourself uncontrollably moved when it arrives. The problem with taxonomy, is that taxonomy is the struggle to *prevent* feeling reverence for what you are describing. The ancient Greeks couldn't figure it out (Sorites). Darwin discovered that he could not begin to categorize the majesty he saw around him (even if he later did). The struggle to associate this thing with that thing struggles against its objective, because if two things share a thing in common, what they share in common will become greater than they. Perhaps the reason humans no longer see God in Nature, is because humans think they can describe God completely? Yet when a child is born, no matter how many times it happens to the human species, humans cannot begin to describe how that novel stimulus makes them feel.

REFERENCES

Arndt, J., Routledge, C., Cox, C. R., & Goldenberg, J. L. (2005). The worm at the core: A terror management perspective on the roots of psychological dysfunction. *Applied & Preventive Psychology*, *11*(3), 191–213. doi:10.1016/j.appsy.2005.07.002

Austen, J. (2019). *Sense and sensibility*. Oxford, UK: Oxford University Press.

Biernat, M., & Danaher, K. (2013). Prejudice. In I. B. Weiner (Ed.), Handbook of psychology (Vol. 5, pp. 341–367). Hoboken, NJ: Wiley.

Bose, K. S. (1996). The theoretical foundations of Zen Buddhism. *The Eastern Buddhist, 29*(1), 85-98. Retrieved from https://www.jstor.org/stable/44362151

Brauser, D. (2017, May 31). Orthorexia nervosa: When 'healthy' eating turns dangerous. *Medscape*.

Brosnan, C., & Michael, M. (2014). Enacting the 'neuro' in practice: Translational research, adhesion and the promise of porosity. *Social Studies of Science, 44*(5), 680–700. doi:10.1177/0306312714534333 PMID:25362829

Ceballos, G., Ehrlich, P. R., Barnosky, A. D., García, A., Pringle, R. M., & Palmer, T. M. (2015). Accelerated modern human–induced species losses: Entering the sixth mass extinction. *Science Advances, 1*(5), e1400253. doi:10.1126ciadv.1400253 PMID:26601195

Cooper, K. B., Gustafson, N., & Salah, J. G. (2013). *Becoming a great school: Harnessing the powers of quality management and collaborative leadership*. Lanham, MD: Rowman & Littlefield Education.

Damasio, A. (2003). *Looking for Spinoza: Joy, sorrow, and the feeling brain*. Orlando, FL: Harcourt.

Darwin, C. R. (1876). *The origin of species by means of natural selection, or the preservation of favoured races in the struggle for life* (6th ed.). London, UK: John Murray.

Davis, G. (2018, August 2). Man and nature: Perspectives on sustainability from east Asian philosophy. *Innovation Management*. Retrieved from https://innovationmanagement.se/

Deleuze, G., Guattari, F., & Foucault, M. (2009). *Anti-Oedipus: Capitalism and schizophrenia*. New York: Penguin.

Dickens, C. (1995). *A Christmas carol and other stories*. New York: Modern Library.

Downing, L. (2013). George Berkeley. In E. N. Zalta (Ed.), *The Stanford Encyclopedia of Philosophy*. Retrieved from https://plato.stanford.edu/entries/berkeley/

Eliot, T. S., & Scheffler, A. (2009). *Old Possum's book of practical cats*. Boston: Houghton Mifflin Harcourt.

Fiske, S. T. (2011). *Envy up, scorn down: How status divides us*. New York: Russell Sage Foundation.

Frankl, V. E. (2006). *Man's search for meaning*. Boston, MA: Beacon Press.

FreudS. (1919). *Das unheimliche* [The uncanny]. Retrieved from http://web.mit.edu/allanmc/www/freud1.pdf

Freud, S. (1920). *A general introduction to psychoanalysis*. New York, NY: Horace Liveright. doi:10.1037/10667-000

Freud, S. (1983). *Totem and taboo*. London: Ark Paperbacks.

Freud, S. (2019). In *Encyclopedia of Death and Dying online*. Retrieved from http://www.deathreference.com/

Gould, K. (1976). The question [Recorded by K. Gould]. On True stories [LP]. Waco, TX: Myrrh Records.

Grandin, T., & Johnson, C. (2005). *Animals in translation: Using the mysteries of autism to decode animal behavior*. New York, NY: Houghton Mifflin Harcourt.

Hergenhahn, B. R., & Henley, T. B. (2014). *Introduction to the history of psychology* (7th ed.). Belmont, CA: Cengage Learning.

Hyde, D. (2011). The sorites paradox. In *Vagueness: A guide* (pp. 1–17). Dordrecht: Springer. doi:10.1007/978-94-007-0375-9_1

Ives, C., Abe, M., & Hick, J. (1992). *Zen Awakening and Society*. London: Palgrave Macmillan Limited. doi:10.1007/978-1-349-11494-8

Karakostas, V. (2012). Realism and objectivism in quantum mechanics. *Journal for General Philosophy of Science*, *43*(1), 1–20. doi:10.100710838-012-9173-5

Keltner, D. (2009). *Born to be good: The science of a meaningful life*. New York, NY: W.W. Norton & Company.

Keltner, D., & Haidt, J. (2003). Approaching awe: A moral, aesthetic, and spiritual emotion. *Cognition and Emotion*, *17*, 297–314. doi:10.1080/02699930302297 PMID:29715721

Ludwig, K., & Ray, G. (2002). Vagueness and the sorites paradox. *Philosophical Perspectives*, *16*, 419–461. Retrieved from https://www.jstor.org/stable/3840922

Meinecke, L. D. (2017). *Neglected by assessment: Industry versus inferiority in the competition for scarce kidneys* (Doctoral dissertation). Available from ProQuest Dissertations and Theses database.

Meinecke, L. D. (2018a, April 14). The uncanny fear of loss: Part 1, Losing the unthinkable. *Psychology Today*. Retrieved on March 8, 2020 from https://www.psychologytoday.com/us/blog/theory-and-praxis/201804/the-uncanny-fear-loss-part-1

Meinecke, L. D. (2018b). A theory of cognitive idolatry, or the human struggle to preserve psychological certainty in a biological species - chapter 1 (Research proposal). doi:10.13140/RG.2.2.16527.43684

Meinecke, L. D. (2018c). A theory of cognitive idolatry, or the human struggle to preserve psychological certainty in a biological species - chapter 2 (Research proposal). doi:10.13140/RG.2.2.31089.53601

Muir, J., & Highland, C. (2001). *Meditations of John Muir: Nature's temple*. Berkeley: Wilderness Press.

Nagatomo, S. (2020). Japanese Zen Buddhist philosophy. In E. N. Zalta (Ed.), *The Stanford Encyclopedia of Philosophy*. Retrieved from https://plato.stanford.edu/

NHS Choices. (2013, August 15). *News analysis: Controversial mental health guide DSM-5*. Retrieved on March 8, 2020 from https://www.nhs.uk/news/2013/08August/Pages/controversy-mental-health-diagnosis-and-treatment-dsm5.aspx

Nonlocality and entanglement. (2019). In *The physics of the universe*. Retrieved from https://www.physicsoftheuniverse.com/topics_quantum_nonlocality.html

Pavlov's dogs. (2019). *Dictionary.com*. Retrieved from https://www.psychologistworld.com/behavior/pavlov-dogs-classical-conditioning

Pawlett, W. (2007). *Jean Baudrillard: Against Banality*. Routledge. doi:10.4324/9780203937365

Pervin, L. A. (1994). A critical analysis of current trait theory. *Psychological Inquiry*, *5*(2), 103–113. doi:10.120715327965pli0502_1

Rosemont, D. A., Sharpe, K., & Bauman, T. (Producers), & Hardy, R. (Director). (2000). *High noon* [Motion picture]. Canada: Artisan.

Shakespeare, W. (2010). *The merchant of Venice* (B. Mowat & P. Werstine, Eds.). New York: Simon & Schuster Paperbacks.

Sherif, M., Harvey, O. J., Hood, W. R., & Sherif, C. W. (1961). *Intergroup conflict and group relations: The robbers cave experiment*. Norman, OK: The Institute of Group Relations.

Spinelli, E. (2013). Relatedness and the therapeutic relationship as viewed by existential therapy. *Dasein, 40*(1), 17-29. Retrieved from http://www.psicoterapiaesistenziale.org/

Spinoza, B., & Curley, E. M. (1994). *A Spinoza reader: The Ethics and other works*. Princeton, NJ: Princeton University Press.

Spurgeon, C. H. (2013). *The salt-cellars: Being a collection of proverbs, together with homely notes thereon*. Harrington, DE: Delmarva Publications.

Sundararajan, L. (2015). *Understanding emotion in Chinese culture*. Cham, Switzerland: Springer. doi:10.1007/978-3-319-18221-6

Taxonomy. (2019). In *Oxford English Dictionary online*. Retrieved from https://en.oxforddictionaries.com/

Taxonomy. (2020). In *Merriam-Webster.com Dictionar.* https://www.merriam-webster.com/dictionary/

Unterseher, M., Westphal, B., Amelang, N., & Jansen, F. (2012). 3,000 species and no end – species richness and community pattern of woodland macrofungi in Mecklenburg-Western Pomerania, Germany. *Mycological Progress, 11*(2), 543–554. doi:10.100711557-011-0769-7

Vrobel, S. (2011). *Fractal time: Why a watched kettle never boils*. River Edge, NJ: World Scientific & Imperial College Press. doi:10.1142/7659

Wilson, E. O. (2016). *Half-earth: Our planet's fight for life*. New York, NY: Liveright Publishing.

Wittgenstein, L. (1965). *The blue and brown books: Preliminary studies for the Philosophical Investigations*. New York, NY: Harper Perennial.

Yamamoto, K., Tanaka, S., Kobayashi, H., Kozima, H., & Hashiya, K. (2009). A non-humanoid robot in the "uncanny valley": Experimental analysis of the reaction to behavioral contingency in 2-3 year old children. *PLoS One, 4*(9), 1–6. doi:10.1371/journal.pone.0006974 PMID:19759818

Zeigarnik, B. (1927). On finished and unfinished tasks. In W. D. Ellis (Ed.), A source book of Gestalt psychology (pp. 300-314). London, UK: Kegan Paul, Trench, Trubner & Company.

Zimbardo, P. G. (2007). The Lucifer effect: Understanding how good people turn evil [Kindle ed.]. New York, NY: Random House.

ENDNOTE

1 Teleological: explanation of phenomena in terms of the purposes they serve under an assumption of predestined human domination of nature, rather than what might have caused their existence whether people came to exist or not.

Chapter 4
Reverence for Childhood and Old Age

ABSTRACT

Chapter 4 draws from one of the authors' dissertations to highlight an injustice some call ageism. The epoch known as childhood is going extinct as the need for more and more education reaches into infancy for the earlier brain growth needed to compete with the velocity of science. The epoch known as old age is undergoing a metamorphosis from something to look forward to, to something to fear. Only the productive middle of an otherwise unproductive lifespan seems to be of value to society. This chapter compares the withholding of regard from children until it's earned, to the withholding of reverence for nature until it perishes. The authors issue a call to honor the very young and the very old, because those are the precious sunrises and sunsets of the human lifespan.

INTRODUCTION

"Hold childhood in reverence, and do not be in any hurry to judge it for good or ill . . .Give nature time to work before you take over her business"
–Jean-Jacques Rousseau

In this chapter, the authors will propose that every moment matters. The authors will use this axiom to explore that missing reverence for each and every moment that humans are fortunate to have. The authors will use this

DOI: 10.4018/978-1-7998-4408-2.ch004

axiom to explore the supposed stages of the lifespan . . . and in the process, propose that *every* stage of the lifespan matters just as much as every other stage. These authors will not uphold the common belief that childhood is here mainly so humans can grow up; these authors will not uphold the idea that old age is here because some humans survived long enough to look back. Instead, the authors of this book about reverence for Life, will hazard that the stage looked upon as "childhood" is just as valuable as the middle and final stages of the human lifespan, whether everybody grows up or not. These authors will venture that it is just as vital not to pity the old for living "too long," as it is not to grieve over a child who dies "too young." If one has lived, surely that was better than not having lived at all? If one does not attain some ideal "stage" or achieve some monumental thing before death, has one's life been lived in vain? No. Viktor Frankl, a famous psychiatrist/survivor of the Holocaust, said no (Frankl, 1990). Life itself is an achievement.

For these reasons and more, every stage of the lifespan matters. Human beings are not here to disparage childhood because children need to start working and earn their keep. Human beings are not here to work until they are too old to work and constantly worry whether they will become an inconvenient burden to their hard-working children. Childhood and old age are not "liabilities." The human species is not here to get a job (Gray, 2013; Meinecke, 2017). It is when adults teach their children they need to "accomplish something or else!" that a perfect child from a perfect family unexpectedly takes her own life (because being a child was not enough to bring her parents joy . . . they expected *more* from her than simply being their child; Meinecke, 2017). In the same way, it is when people expect to live an entire year and live just nine months, that they forget how grateful they were for each and every day.

The reader will recall that this was the premise in the last chapter, too, when the authors proposed that every other species is not here for the sake of the human species. Nor is any member of any species expendable in order to protect or preserve the idea of its species from perishing. Recall that a "species" is an idea, whereas an individual being is not an idea (until it dies). Every individual matters. Humans have produced memorable films emphasizing this idea (Ford, 1945). There is no one exactly like that one unique being in all the world, and never will be again. In this is reverence (for Life). In this is love (of Life). In this simple understanding of the value of the least of things, humans may (perhaps) save the greatest of things (this beloved planet) from an early death. It is not the human habit of putting things into categories that makes each member of that category seem more indispensable, but *less* indispensable (Meinecke, 2017). It is when people as a whole fail to completely see the value of an individual person, that the missing value of that individual becomes the apparent significance of the category which describes that missing regard (Marchant, 1646; Meinecke, 2017; Rogers, 1961; Spinoza, 2012). But when people as a whole see an individual person or thing completely, they abandon all else to discover see how endlessly astonishing that person or thing really is—"no matter how hopeless, no matter how far" (Darion & Leigh, 1966; Matt. 13:46; Spinoza, 2012).

EVERY MOMENT MATTERS

Why don't humans treasure every moment of their lives? That is a very good question. If one moment is good, aren't two moments better? Curiously, no. And even curiouser, humans never seem to live just one moment—well, if anyone ever did, no one can remember. You cannot compare a moment to itself. And if humans live just two moments, humans always struggle to compare them to see which moment was better, or which moment humans felt *more* alive. More alive? What does that mean? Isn't being alive better than the alternative? Ask yourself this question: If you have lived even one moment, isn't that better than never having lived? Most people would say that one moment is not enough. Most people would compare how much time they get to how much time everybody else got, since how else can one gauge the human lifespan but by the average? Yet when each moment is compared to the average, each and every moment doesn't matter anymore. Now, unless life lasts as long "the average," humans worry they haven't really lived. But if there were no average, no one would compare what they got to that average.

If humans do not live "long enough" humans grieve. If humans do not live as long as *that* guy, they grieve. If *that* guy does not live as long as *this* guy, humans grieve. Only those who have never lived, never grieve. If humans live two moments, they compare them to see which moment was better, just as they compare being alive to never having lived. What if there were only one moment at a time, and no other moment existed until this moment had been perceived? Would that mean every moment mattered?

EVERY STAGE OF LIFE IS THE BEST STAGE

Isn't it funny how nothing stays the same, yet humans struggle to perceive it that way? Take insects, for instance. On one day the same insect is "an egg." On the next day that insect is "a pupa." Then it is a "larva." Lastly, the egg is an "adult." How can an infant be an adult? Well, if you look at the adult *too soon*, it is an infant. If you look at the infant *too late*, it is an adult. It is kind of like the dual-slit experiment using time-lapse photography and a very bad memory. Now it's a particle; now it's a wave. It's whatever you want it to be, if it has to be the same thing—no matter when you look. But what if each time *anybody* looks, each thing is completely different? What if in order to see a distinction, it must not be like anything people have ever seen before?

The human lifespan is quite long—about 100 years, give or take a few decades. So humans divvy it up into incredibly different looking physical and mental "stages" (like infancy, childhood, adolescence, middle age, and old age). The lifespan of grief is quite long too. So humans divvy grief up into stages as well (denial, anger, bargaining, helplessness, and "acceptance" according to Kübler-Ross & Kessler, 2005). Then humans argue over how long humans ought to grieve before it's considered a disorder, or humans accuse people who haven't grieved "long enough"[1]. Even though every species has a different number of years they're expected to live, some longer and some very brief, humans have found a way to chunk them into something less natural than instants, even if instants are what all the species share, and all those who grieve share too. Humans are in the egg. Humans are outside the egg. Humans are just a pupa. Humans are finally a butterfly. Humans are in the grave. And even though every stage is a stage of the same life, humans tend to think of only one stage as the purpose of all the others anyway, just as humans think the purpose of a population of things is its average or ideal (whatever that is).

So, even if humans have lived inside an egg, humans cry because they didn't break out of it. Even if humans have grazed like happy caterpillars all summer, they cry because they didn't turn into a butterfly before they died. And even when they've reached a ripe old age, can look back on a lifetime of life in the egg, can look back on a lifetime of life crawling happily around, can look back on a lifetime of flitting here and there (before their wings began to fray and had to crawl once more), they cry as they look back anyway. Humans cry because they are no longer able to create something to look back on. Or, they cry because there was one stage they miss most of all—as though everything before and after that stage were here just for that one stage of the lifespan. But living in an egg is having lived. Creeping as a caterpillar is having crept. Flitting as a butterfly (having flitted) was not the only time one lived, nor is it true that—if one should die before one becomes a butterfly—one hasn't lived. Dying before growing up does not mean life was lived in vain. Dying during adolescence does not mean all that parental investment during infancy was in vain. Only butterflies miss being caterpillars. Caterpillars do not miss being caterpillars (unless they are taught that caterpillars are here to "become" butterflies; then and only then may they realize they will never make it that far, and start to cry). Every stage of the human lifespan matters just as much as every other. This seems true no matter which "species" a creature thinks it belongs to.

BEING IS BETTER THAN BECOMING

Lots and lots of books talk about the benefits of becoming. Not so many books talk about the joy of being something right now. If all humans do is look forward to what humans will become, when will humans look back? And if all humans do is look forward to looking back, shouldn't it be enough to become something wonderful for an instant, so humans can spend the first half of their lives looking forward to that instant, and the second half looking back on that instant? Of course not; what good is becoming something, if most of the time you are something else?

Being is better than becoming. The hope of being is better than the hope of becoming. That's because *being* always happens, whereas *becoming* may or may not happen. Whatever a child turns out to be, a child cannot fail to be a child. It is only when grownups tell a child he or she is not yet an adult, and that he or she better hurry up and act like one, that a child worries about the future. This effect has been demonstrated by a French psychologist named Phillippe Rochat (2003). Between the ages of two and four, on average, a child stops reasoning like a child, and begins to reason like a man. Rochat pasted a simple sticky note to the foreheads of children. Younger children rejoiced in the silliness and the stares they got. "Look at me!" Older children became self-conscious. "I'm sorry, don't look at me!" Maybe younger children don't worry about becoming an adult. Maybe they are happy to be alive. But as they grow up, children worry more and more about whether they will grow up, and whether they will grow up "right." But how can they fail to be children, if they already are children? Grownups make them feel guilty for being children. Why? Grownups make them feel guilty because they look at childhood as a stage on the way to becoming an adult—as though the *imago*[2] stage of species hominin were the be-all and end-all of being human in the first place, and the other stages steal from the *imago* (Imago, 2020).

Otherwise, the concept of childhood is an everlasting Day, an eternal stage with no beginning or end, nor any form or function except "to play." Playing is being; the epoch of childhood is the kingdom of play (Gray, 2013). Working is not being (because working is looking forward to playing—earning enough to spend your last few years having fun). Becoming is not playing. Becoming is way too serious, way too risky to be a game with no winners or losers. Only some children win. Most children may end up in therapy, trying to figure out what went wrong ("why didn't I grow up?"). They feel guilty for being alive. Some try and fix that.

But Darwin didn't look at living things this way. Darwin didn't gather a bunch of leaves to see which ones were being leaves correctly and which ones needed therapy. Instead, he noted that *each and every leaf* on the same tree was utterly *unlike any other leaf* . . . and yet they all seemed to be similar somehow (Darwin, 1876). Aren't leaves a lot like children? To modern humans, growing up is more like a competition than enjoying the age you are at. The problem with that is, where do you draw the line between "healthy competition" and plain old neurosis? (Ryckman, Thornton, & Butler, 1994). Competition requires a lot of comparison, such as "which of us leaves on this same tree is most like an ideal leaf?" or "which of us children from this same species is most like an ideal child?" It is odd that the disciples asked Jesus much the same question, which seemed to upset Him:

Then a dispute arose among them as to which of them would be greatest. And Jesus, perceiving the thought of their heart, took a little child and set him by Him, and said to them, "Whoever receives this little child in My name receives Me; and whoever receives Me receives Him who sent Me. For he who is least among you all will be great. (Luke 9:46-48)

Did Jesus say that children had to grow up and become disciples in order to please God? No, He told His disciples to act more like children, not more like disciples. He didn't tell everybody to become His disciples, or make an industry out of it. He said this on more than one occasion, for example, when the disciples tried to shoo away the children who came to touch Jesus:

Then were there brought unto him little children, that he should put his hands on them, and pray: and the disciples rebuked them. But Jesus said, Suffer little children, and forbid them not, to come unto me: for of such is the kingdom of heaven. (Matt. 19:13-14)

If a child must "become" a grownup, a child might not become a grownup. But if a child does not need to "become" anything, such a child cannot fail to be. Parents that let their children know how grateful they are simply to have them, will not make their children feel worthless. But many parents make it plain to their children how upset they are at them for not growing up quicker. Not all children will become successful. Not all children will thrive. Sadly, children who fail to thrive, who fail to "grow up" and become "an *imago* like everybody else," grieve as though they had *never lived.* And adults who did grow up (but didn't seem to thrive before they became too old to be useful), grieve as though they never grew up! (Obviously, they did). These concepts are famously engraved in the *grand theories* of psychology . . . such as those by Freud, and Erikson, and Piaget (Belsky, 2010). Why do children weep, they asked? What went wrong with the raising of those children? How can grownups "fix" them?

This book about Biophilia is about revering life no matter what it looks like, and no matter whether it "grows up" or not. This book is about reverence for Life in all its kinds, great or small, child or beast, immature infant, or senile old man (Marchant, 1646; Meyer & Bergel, 2012). This chapter contends that people often weep because society didn't give them sufficient reason to be glad before they died. Humans spend their childhood thinking they are too young. Humans spend their golden years thinking they are too old. And they spend the middle fighting for a sparse respect that is rarely and reluctantly given.

One of the authors asked the question of why humans grieve in his dissertation, but within the framework of people facing death when there is a chance for life (Meinecke, 2017). He wrote about the minority groups who have to wait too long to be saved, while other groups with more to offer do not have to wait very long. He noticed that it's nobody's fault really, since only random selection seems to be just and fair when absolutely everybody needs the same thing, but there isn't enough to go around. He noticed that the pursuit of perfect health and endless wealth are bad things, because if everybody is healthy and wealthy, nobody stands out. If nobody stands out, there is no way to decide which ones should receive a scarce kidney transplant to replace a failing kidney. He noticed that, if only society will give you what you need to stay alive before it's too late, you won't have to grieve very long, and neither will your loved ones. But some people are Too Young to get what they need right now, and may not grow up to be of much use to society, while other people are Too Old right now, and are unlikely to benefit society anymore. Even if society could give them what they worked so hard for, they are too old and too sick to enjoy it. And some people are Just Right, but there are just too many of them to receive what society has so little of (because now *too many people* are healthy). So, even though people are every bit as alive during all three of life's stages (childhood, middle age, and old age), only people who are Just the Right Age are completely valued by society. The others are all here for that stage.

Logically, every stage can't be the "favorite" stage, can it? Otherwise, why divide them up at all? Everybody can't be above average, nor can every stage of the lifespan be better than every other (Sharot, Korn, & Dolan, 2011). And so, humans view the human lifespan as though they *will be* human one day (childhood), or *used to be* human (old age), or *are* temporarily human (because they are temporarily useful to society; Meinecke, 2017). For human society, it's the age from 18 to 65 that is designated as fully human. It just so happens to coincide with the age people are fully employable. But childhood (under 18) and old age (over 65) deserve to be treated as fully human too—just as minorities and non-citizens who are the perfect age to be employed deserve to be treated as fully human too. This chapter is about how Society denies the young and the old *unconditional positive regard*, because Society is too busy praising the productive years for all that wonderful productivity (Rogers, 1961). The outcome is not pretty: children worry that Society will never love them, and seniors worry that Society will never love them again.

REVERENCE AND THE COMPETITION FOR SPARSE REGARD

There is a sort of unspoken competition for reverence among the human species, as though positive regard were a sparse commodity. There never seems to be enough reverence to go around. Sound familiar? The famous Carl Rogers (1961) proposed that maybe there is more than enough *unconditional positive regard* to go around, but for some reason, the human species would rather grant it reluctantly anyway. Humans like to save their reverence for special occasions, instead of practicing universal reverence for every living thing they see, every day of the week. So, only the fittest (most productive) merit human reverence—even if they probably don't need as much reassurance as those who would give anything to be regarded even a little bit well (Meinecke, 2017).

Yet positive regard doesn't cost humans anything. Telling someone you appreciate them doesn't take a lot of time or a lot of energy. This species could easily smile adoringly at every person it met on the street, like a bottomless basket of Duchenne smiles, and still get to its appointments on time—and still feign that *social smile* its been saving up for just that *schadenfreude* occasion. This species could easily cheer every time someone performed for it, no matter how great they were (or not!), and still have plenty of applause left over for its "favorite" performers. But this species doesn't. No, humans pass right by dozens of wishful people silently—on their way to their solemn appointments—as though saving their sparse regard for the one specific person they are going to see. Humans hold back their praise six days a week so they can have enough for Sunday. They save their applause for that one performance they all think is just right, and only then do they release their laud—a *pseudobulbar* response, eerily reminiscent of the bitter quanta of their reluctance to clap up to that point. Why do humans do that? Why are humans so stingy with their praise? Their children come to them with a scribbly card and say, "Here Mommy, here Daddy, I made this for you!" They look at the card and scowl, and unexpectedly direct their children to go practice their handwriting.

GOLDILOCKS AND THE THREE BEARS

Reverence is a lot like the story of *Goldilocks and the Three Bears*. Why do the authors suggest this? Well, one of the authors of this book found in his dissertation, that humans reserve their reverence for those who are "just right" (Meinecke, 2017). Remember Goldilocks? When she came upon porridge or chairs or beds, some were too this, some were too that, and some were just right. As it turns out, little girls looking for food and a place to sleep aren't the only ones who look at their choices as too hot or too cold, too hard or too soft, or the option that is just right (even when humans are simply looking for any amount of food and rest). Because when human candidates are looking for a replacement kidney to save their lives, they need to register for a list that prioritizes who gets a kidney first. They do that because there are far too many people in need of a kidney, and nowhere near enough kidneys. As of the date of the data collection of the author's dissertation, for example, there were around 678,000 people who needed a kidney to survive. Out of that number, only 88,000 made the kidney transplant waiting list. And out of that number, just under 18,000 actually received a kidney. Most people perish, unfortunately, while waiting for their turn. But somebody has to decide who gets a kidney, and somebody has to decide who has to wait. It's a very difficult process for everybody concerned.

The author of the study divided Physical Age into three groups—Too Young (0-17), Too Old (65 and over), and Just Right (18 – 64). Just so you know, Physical Age was the chronological age of a candidate regardless of physical or mental condition. By splitting the candidates on the transplant waiting list into three groups, (too young, too old, and just right), the data analysis was able to reveal some interesting things about sparse positive regard. Despite never having contributed to society, the most mentally fit children had to wait the least (Meinecke, 2017). Despite having contributed a lifetime of service to society, nearly all seniors had to wait the most. But the most surprising finding, was that people whose age was "just right" had the worst chance of getting a kidney—because there were just too many candidates who were "just right".

HONOR CHILDHOOD

The part of the lifespan called Too Young will be discussed first. What does that mean—too young for what? Too young for Life? Remember Goldilocks? That little girl had not one but three bowls of porridge to eat. But one was Too Hot. And the other one was Too Cold. Somehow, even when you are a hungry little girl lost in a forest, you can afford to be picky! In the authors' research, when you need to be picky, the authors discovered that the age of industry (between 18 and 65) is typically just right. Anybody under 18 is too young, and anybody over 65 is too old. Only people in the middle are People. The others are viewed as living liabilities (unless they grow up to be People, even if old people can't grow up anymore). How old do humans have to be, to be human? Aren't humans human when humans are born? Maybe even before that? So, when is a person a Person—only when a person is "just right?" (Maybe it's simply when People love them?). Some authors have taken this in really amazing directions, such as pointing out the human habit of dehumanizing anything that isn't just like them (Hodson, MacInnis, & Costello, 2014). It's funny how affection and judgment are so closely and conditionally related (Fiske & North, 2014).

When the author of the waiting list research looked deeply into this epoch under the arbitrary age of 18 in physical years, he noticed that some were *mentally* younger than they seemed physically, and so, though they had waited the appropriate amount of time, and though their bodies were physically mature, they would never be called people—because their brains were not as developed as the brains in the just right group. He also noticed that some were far more mature, mentally, but penalized for being too young physically, or given organs that those with less intelligence needed but could never earn because they would never be as smart. And the most bizarre finding, was "This study concluded there was an inadvertent irony in candidate assessment criteria. The need to ethically allocate scarce organs has resulted in criteria that favor the survival of human kidneys over human candidates" (Meinecke, 2017, para. 1).

What does it mean to honor Life, if in the end civilized society has to prioritize the survival of scarce organs over human candidates—because civilized society cherishes life so much? Far more people need a transplant than there are transplants. And far more children crave civilization's affection than the few who get most of it (because they grew up "perfect" while other children seem "impaired"). What happens to the child that did everything he or she could, but still doesn't get civilization's applause? And why does civilized society treat the first part of the lifespan as "too young?" Too young for regard? Like a little girl looking for some porridge, does a parent need to have a child who is "just right" to be grateful for, and to simply cherish that child? Treating the human lifespan as though it were divided into three groups, is like making children wait for a chance to receive a scarce and lifesaving organ. Childhood has become a waiting list of candidates hoping to be given a label society calls *industrious*. Most of them, sadly, receive a label called *inferior*, and perish deep inside (even if their outside is fine). But society could honor childhood no matter its outcome, and no child's spirit would perish before its time (Meinecke, 2017).

There is a pronounced effort in the world today, to steal childhood away from children (Gray, 2013). Has the reader noticed? The demand for the mental development of human children is reaching into the stages of their physical development, until the simple idea of playtime prior to the age of human industry seems like a human infant's unforgivable sin (Kivnick & Wells, 2014; Rochat, 2003; Stone, Crooks, & Owen, 2013). Not long ago, society held a deep reverence for childhood (and old age). Society watched its toddlers grow, blissfully unaware of the raging epoch of industry and productivity all about them, wherein grown-ups burn their mortal candles out before their time. Society used to long to be young again, if only to forget what it's like to be "mature." But now if your toddler isn't in preschool by the time that child can crawl, both you and your child may find yourselves the objects of neighborhood gossip and scorn. Not long ago, Society scorned its children only because they had failed to become bipedal by the age Society deemed just right to walk upright. Now Society scorns its children if they walk upright too often, when they ought to sit down again so they can learn their letters. Childhood has become prep school, an infant academy, instead of letting the inhabitants of this rare period of the lifespan revel in their time. As the author wrote in his dissertation:

Children are being denied play, compelled to study serious subjects at earlier ages, and curiously experiencing novel disorders such as ADHD, depression, autism, and learned helplessness at the dawn of life (Panksepp, 2007). As evidenced in this study, physically developing children may be denied the right to Life simply because their brains and thought processes are not adequately developed, even if their bodies remain biologically viable and in need of immediate (not promised) warmth (James 2:15-16). Like the human view that fish do not feel pain (so the industry of fishing may flourish without regret), it is possible that society does not think children feel pain when told to ignore their childhood—so the industry of informational reverence may flourish without regret, while the species of play goes extinct. (Meinecke, 2017, p. 173)

And so an application of his research, which he envisioned, was to permit play while humans are young enough to play. It is not fair to reach into childhood because the just right group cannot get enough workers to work early enough to compete globally with everybody else's children. Childhood is an epoch of life unlike any other, so if humans have reverence for Life, humans ought to revere Childhood too. As Schweitzer envisioned such a reverence, "All of life is equally sacred. No life is second or third class—all life is equally valuable" (Meyer & Bergel, 2012, p. xiii). Moreover, no part of the lifespan is more like Nature's innocence and wonder than when humans are children. It's a time when awe for Life is an *instinct*, and great human achievements pale in comparison to a buzzing bee kissing a blossoming flower on a blustery day.

Let the authors share a secret with you (*voici notre secret)*. One of the authors lived in a wooded community where nearly everything was natural and alive. Driving around got quite confusing sometimes, though, searching for homes with garage sales. But he and his wife stumbled on an oil painting one afternoon, and he bought it. Like some Hollywood movie mystery, after he got it home, he began to research the artist who made it, because he was a grad student and that's what grad students do. Lo and behold the artist was actually a famous educator at the turn of the 20th century—Preston Willis Search. And the author of the preface to one of Search's prominent works, was none other than G. Stanley Hall—noted psychologist. Just so you know, G. Stanley Hall translated Freud. Now you're in-the-know! The volume was all about honoring childhood and children—not just education and success. Since the author loved his gift of education, he was intrigued. What could be better than being educated? Well, he read that humans seem to have forgotten, in just the space of a hundred years, what life was like before all human children spent the first third of their lifespan indoors, learning about life through darkened windows called books—instead of playing outside and learning about life in the sunlight from living things (like curious feral animals on the way to school, and small curious insects in an old widow's flower bed up the street). A hundred years ago G. Stanley Hall noticed how childhood was already dying, and he noticed it enough to write about it—in Search (1901):

When a child begins to go to school the change of his environment is very great. Instead of constant activity, he must now sit still and keep still; instead of moving his hands and arms freely, the strain of effort is now focussed [sic] upon the very few tiny, pen-wagging muscles. The eyes, instead of moving freely, are confined in the zigzag treadmill of the printed line. It is no wonder, therefore, that the child so commonly loses weight on first entering school; that short-sightedness and other eye troubles increase almost regularly through the school period; that headaches, anaemia, scoliosis, defects in development if not signs of disease, appear in the stomach, heart, and lungs, and especially in the nervous system, the gradual deterioration of which is hard to recognize. (p. 55-56)

Humans seem to have forgotten what they've done to their children. They seem to have forgotten what they've done to their planet. They seem to have forgotten how to be reverent, and what to be reverent of. So, what kinds of life should humans have reverence for? Should humans only revere living things aged 18 to 65? Is childhood here so humans can revere adulthood? Or is childhood already a thing, with a beginning and a middle and an end? Maybe childhood is a species, and maybe that species survives to remind humanity what the planet was like before humanity made it stay outdoors while it locked its children inside. Maybe children aren't naïve, just innocent. Maybe a child remembers what Nature was like before humans conquered it, tamed it, schooled it, filled it with facts instead of affection, and domesticated the awe of juvenility away?

If humans truly honor life, humans should honor life in all its kinds, by honoring childhood too (because children seem more alive than most humans do). Biophilia isn't just about honoring the imago, because humans can honor the egg, the larva, and the pupa too. All of them are alive, but some of them are still invisible to the eye (like the fox told the Little Prince, when he was sharing his secret; Saint-Exupéry, 1943). Reverence for life is about about reverence for juvenility too. Yet so curiously, when one of the authors did his exhaustive literature review for chapter two of his dissertation, he found an eerie similarity between the scorn for the juvenility of things humans regard as crops, and a scorn for the juvenility of things humans regard as children. Humans see their children as unripened fruit that need to grow up so they can harvest them. Oddly, that's actually not what early religious scholars thought about Life. Jacques Marchant (1646), while trying to describe the idea of God, once penned the following lovely phrase: "Fugit infantia, fugit pueritia, fugit adolescentia, fugit juvenius, fugit senectus: et hodie non est, qui heri fuit." What's that you say—Latin? Yes, it's Latin. People used to write in Latin because religious scholars around the world spoke the same language—much as physicians and medical scholars use Latin today, and for the same reason. What does it mean? It means: "Infancy flees, childhood flees, adolescence flees, youth flees, old age flees: and nothing is here today that left yesterday" (p. 129). It means that nothing stays the same except God. Yet everybody misses those things even more than everybody misses God Himself—why is that? Maybe it's because the loss of things everybody cherished is how God makes Himself plain enough to see; just maybe that's why children are said to be more like the inhabitants of heaven? (Matt. 19:14). Whatever is briefer helps a person take in the volume and value of these moments. So maybe when humans fail to revere childhood, it isn't just children humans fail to revere.

A PRAYER FOR OLD AGE

It's time to talk about the part of the lifespan called "Too Old" now. What does that mean—too old to be alive? Why are certain ages of the lifespan ideal? Isn't being alive ideal? And why are those years not the golden years? Why are people so disabled and worn out by the time they reach the "best years" of their lives? As the authors mentioned, in one of the authors' dissertations Physical Age was divided into three groups—Too Young (0-17), Too Old (65 and over), and Just Right (18 – 64). Now add the fact that Mental Age was also divided into three groups—Too Slow, Somewhat Slow, and Just Right. Just so you know, Mental age was the numerator of the Intelligence Quotient (IQ), which is an official assessment of how mentally developed a person is relative to the norm for that chronological age. Phew.

Hey! How come there isn't a Too Dumb, Too Smart, and Just Right? That's because, unlike physical age differences (too young or too old), mental ability does not seem to have a "too smart" group, only several "not smart enough" groups. People don't normally say someone is Too Intelligent to be useful, not the way people call someone Too Young or Too Old to be useful. Like the physical world, people seem to view the body as either not useful enough yet (too young), useful (just right), or no longer useful (too old). But the mind is different; people seem to view the mind as not useful at all (too slow), almost useful enough (somewhat slow), or useful (just right). There is a lot of favoritism for the human intellect at any age these days (and intellect is not at all like Nature) and an even more obvious scorn for the human body at any age (and the body *is* a lot like Nature; Meinecke, 2018a, 2018b).

Awe for Life, though, unlike awe for Intellect, is not about reserving reverence for the "productive years," while the less productive years are more like liabilities. Every year of physical life is precious, and not one day is less precious than another. Notwithstanding this fact, many seniors think back on a life they will never live again, wondering if they lived it well. All they have left is a kind of lottery, and the hope they will receive some warmth from society, if any warmth is left over for old age (Fiske, 2013). The famous psychiatrist Viktor Frankl once helped a woman find her worth in old age, by reminding her that no life is lived in vain (Frankl, 1990). Yet nowadays, millions of children think their tiny lives are in vain, and are already on antidepressants before they reach the age of industry, or have anything to regret having done. Humans have a label too, called "failure to thrive" that they brand their "underproductive" children with. Such bad habits by grown-ups are not the best examples for their children, nor do they say much about Society's regard for Life. And when those children grow old, they feel their lives were lived in vain.

But whether a human child grows up to be a successful politician or a career scientist does not make the child any less a human being. And how tragic it is that millions of seniors eke out their golden years being told they are "in the way." Is this reverence for old age? Must a man be successful to be treated as a man? Must a woman have a career, to be cherished by men? Life matters (Vazquez, 2013), but when humans get too old to contribute, most societies tell them they don't matter anymore. Long ago, though, even to reach "old age" earned you much respect from your society. The young looked forward to a long life—when they were reminded what to look forward to by a few revered but unproductive elders—who had survived the endless age of industry (Gopnik, 2016).

The author whose dissertation was about the comparison of ages, suggested in the practical implications of his research, to "honor people, not industries of people." All living things are alive, but not all living things are revered. All human beings are alive, but not all human beings are told just how much they mean to society. But Biophilia is about the love of living things—not just successful living things. Reverence for Life is about telling everybody they matter, not some people they matter (Vazquez, 2013). Yet near the conclusion of the study, the author penned:

Recent evidence and this research study suggest that the survival of industry is often prioritized over other concerns, as though the survival of human ideas were more crucial to humans than the survival of human beings (Flacco et al., 2015). This emphasis on industry and competition cannot fail to have an impact on living but impaired individuals struggling to find a niche in the social heart of their culture. These are individuals who crave warmth but need competence to reach it, and who dream of being admired but are more likely to endure scorn (Fiske, 2013). The literature is permeated with case studies and statistics of neglected children (too young), of overworked adults (just right), and of the abandoned elderly (too old), who end up unemployed, uninsured, and unloved—when they somehow failed to achieve industry, initiative, and generativity—though all of them were physically, desperately here. (Meinecke, 2017, p. 175)

Just as anybody can appreciate a wild rose in a mountain meadow, or the sunlight on a wilderness lake, and not worry whether they deserve human regard, so too anybody can appreciate a person just because they are here on this Earth along with everybody else. Old people are still people, and they love to tell the young what they saw of Life, when they too were young. That should be enough reason to honor elders (and the means by which to honor them too)—to spare a little time sitting on a park bench with an old widow or widower, and ask if she or he wouldn't mind sharing their story. As the sun goes down on that conversation, another kind of light seems to dawn on them, like the flame of a reverent candle above their heads—whether the two meet again in this life or not.

HONORING THE ELEMENT OF AWE

As the authors wrap up this chapter on the importance of reverence for the disregarded epochs of the human lifespan, perhaps the reader can identify the commonality between chapters three and four? What is it that one can take away? Chapter three was about how putting unique things into categories seems to steal away the worth of each individual. Chapter four was about remembering to treasure the unique but underappreciated spans of the human lifespan, because that steals away our favorite years. Why are these similar? Perhaps they are similar, because each is about a thing that seems unimportant until there is lots of it (or just enough). But each thing in life is just as important as every other, just as every human child is precious, even if one happens to become a famous astronaut and the others become not-so-famous short-order cooks. Maybe it's the *element of reverence* that determines how much reverence you are able to eventually express? Because, when one can see how precious the least parts of this lifespan are, how can one fail to revere the best? And when every member of a category is in its own category, how can one not revere the category even more? Maybe the element of awe determines how much awe for Life humans end up feeling? Maybe when the smallest thing is overwhelming, it is impossible not to be overwhelmed by the greatest of things?

The difficulty in noticing the elemental aspect of reverence, is that now there are just too many things to revere. Remember the transplant candidate waiting list? What if every candidate were precious, and no one could decide? Who would get a lifesaving kidney then? And what if someone stumbled onto a meadow, and noticed every individual wildflower the meadow held? It would take that someone all day just to get from one end to the other, because that someone would have to stop and adore each and every flower, one by one. As Tolstoy (2019) might have proposed it, how much reverence does a man need? And so, humans probably exhibit favoritism because there are just too many things in Life to adore. Maybe humans aren't stingy after all, maybe the alternative is overwhelming grief at what humans couldn't get to?

Biophilia—reverence for Life—exhibits a kind of irony. If everybody stopped to honor every living thing they saw, they wouldn't get very far in

Life. They would step outside and spend the whole day greeting flowers and insects and birds and mice *in their own backyard*, and never make it to school to greet their teachers and friends (one of the authors actually did that as a child, and spent a good deal of time in the corner for it, too). Or they would step outside after dark (where their teachers said the wild things are!) and end up spending the whole night telling each and every star how lovely it is—and never get any sleep. Reverence for Life is about putting everything around you first, and realizing you may never get around to yourself.

Reverence for Life is like an unfulfillable demand, but you fulfill it anyway because you just can't help it. And so tears squeeze out from between your criticisms, as if to circumvent the very idea of judgment, and you let your heart tell the world "I love you!" (no matter how inhibited your mind insists you should be). That's the kind of reverence developing children *need* to see in us when they are wondering what they ought to revere. That way, if they ever wonder if we love them, they will realize that we love things much simpler than they, and oh how much more we must love them too. And that's the kind of reverence our retired neighbors need to see in us when they are wondering if their lives matter. That way, when they see how much we treasure things that never worked a day in their lives, oh how much more we must treasure them for all they did for us. In treasuring the element of awe, no matter the amount of reverence needed to find awe, no matter what combination of things needed before everyone overflows with awe, the thing everybody reveres will know how much more everybody adores it than they can ever say. When humans express awe, they *exchange* awe, just like they exchange honor with Life itself. And nobody has to go without, because awe isn't a thing everybody can run out of.

There is a concept in psychology called *narcissism*, and it functions much the same way as the hoarding of reverence and honor (Freud, 1920). But really it is just the incapacity to believe others appreciate you, not narcissism. It is the inability to tell if others revere you. In the same way, if nobody hesitated or withheld honor from anybody, nobody would wonder if somebody honored them, and nobody would seem narcissistic just because they were unsure if somebody else needed a little honor. Can the reader see how this is happening to human children? Society calls it *depression*, not narcissism, when children cannot tell if society thinks they have worth. Can the reader see how this is happening in the elderly? Society calls it *grief*, not narcissism, when the elderly do not have a lifespan ahead of them to find out whether society thinks their lives are now lived in vain (Frankl, 1990). And in the middle years, when people have the best years of their lives in their grasp, if people cannot tell whether their friends honor them, their self-doubts comes out as *narcissism*, and their friends mistake their self-doubt for conceit. Freud had this lovely pair of words to describe why some have too much and some have too little. He called them more *manifest* versus more *latent*. Nothing is really hidden or delayed in Nature; it just takes a while for humans to see things humans don't believe in (because humans see the things they believe in right away, even when they aren't there). Favoritism for certain things redistributes the velocity of reverence, so that some things (things that have gone without reverence for a long time) have to wait much too long for affirmation of worth (Meinecke, 2018a, 2018b). So even if "certain things" are certain that humans adore them, far more things become uncertain of whether humans adore them—so that a few will know for sure. Oxytocin studies seem to bear this out.

But with Biophilia, nothing has to go very long before it hears someone say "I love you! I adore you!" With a worldview like Biophilia, a love of every living thing, very few will end up depressed, or sad, or bitter, or vain, just so that a few will know for sure that they are loved. "Simon, son of Jonah, do you love Me?" (John 21:16). One can almost hear the reply: "Then tell Me. Tell Me every day." There is a famous line from way back: "For those who honor Me I will honor" (I Sam: 2:30). One can almost hear the Earth whisper, "Do you honor Me? Then honor your own children. Honor your old and dying children. Honor everything." When everybody honors everything, nothing wonders if it mattered.

CONCLUSION

Childhood is no less valuable than adulthood. Old age need not be spent looking back at Life. Some things are not better than other things; humans simply cannot bear the idea that there is too much to adore before they—and this Life they cherish—part ways forever. So if humans don't hesitate to tell the myriad of "lesser" things (at least once), what humans long to tell the greater over and over, humans will not fail to tell the loveliest things, what humans told the "average" things every single day. Because if humans said it to the least of things (the elements), surely what all things in this world are made of heard it too.

Then the righteous will answer him, 'Lord, when did we see you hungry and feed you, or thirsty and give you something to drink? When did we see you a stranger and invite you in, or needing clothes and clothe you? When did we see you sick or in prison and go to visit you?' The King will reply, 'Truly I tell you, whatever you did for one of the least of these brothers and sisters of mine, you did for me.' (Matthew 25: 37-40)

REFERENCES

Belsky, J. (2010). *Experiencing the life span* (2nd ed.). Worth Publishing.

Darion, J., & Leigh, M. (1966). The impossible dream (the quest) [Recorded by J. Nabors]. On Gomer Pyle, USMC [Television series]. New York: CBS. (November 1967)

Darwin, C. R. (1876). *The origin of species by means of natural selection, or the preservation of favoured races in the struggle for life* (6th ed.). London, UK: John Murray.

Fiske, S. T. (2013). Divided by status: Upward envy and downward scorn. *Proceedings of the American Philosophical Society, 157*(3), 261–268. Retrieved from https://amphilsoc.org/ PMID:25638826

Fiske, S. T., & North, M. S. (2014). Measures of stereotyping and prejudice: Barometers of bias. In G. J. Boyle, D. H. Saklofske, & G. Matthews (Eds.), *Measures of Personality and Social Psychological Constructs* (pp. 684–718). doi:10.1016/B978-0-12-386915-9.00027-9

Flacco, M. E., Manzoli, L., Boccia, S., Capasso, L., Aleksovska, K., Rosso, A., ... Ioannidis, J. P. (2015). Head-to-head randomized trials are mostly industry sponsored and almost always favor the industry sponsor. *Journal of Clinical Epidemiology, 68*(7), 811–820. doi:10.1016/j.jclinepi.2014.12.016 PMID:25748073

Ford, J. (Producer/Director). (1945). *They were expendable* [Motion picture]. USA: Metro-Goldwyn-Mayer.

Frankl, V. (1990). Facing the transitoriness of human existence. *Generations (San Francisco, Calif.), 14*(4), 7. Retrieved from https://www.asaging.org/generations-journal-american-society-aging

Freud, S. (1920). A general theory of the neuroses (G. Stanley Hall, Trans.). In *A general introduction to psychoanalysis* (pp. 10–62). New York, NY: Horace Liveright. doi:10.1037/10667-002

Gopnik, A. (2016). *The gardener and the carpenter: What the new science of child development tells us about the relationship between parents and children*. New York, NY: Farrar, Straus, and Giroux.

Gray, P. (2013, September 18). *The play deficit*. Retrieved from https://aeon.co/magazine/being-human/children-today-are-suffering-a-severe-deficit-of-play/

Haikonen, P. O. (2016). On the sentience of fish. *Animal Sentience: An Interdisciplinary Journal on Animal Feeling, 3*(8), 1-2. Retrieved from https://animalstudiesrepository.org/

Hodson, G., MacInnis, C. C., & Costello, K. (2014). (Over)valuing "humanness" as an aggravator of intergroup prejudices and discrimination. In P. G. Bain, J. Vaes, & J. P. Leyens (Eds.), *Humanness and Dehumanization* (pp. 86–110). New York, NY: Psychology Press.

Imago. (2020). In *Oxford Learner's Dictionaries*. Retrieved from https://www.oxfordlearnersdictionaries.com/us/definition/american_english/imago

Kivnick, H. Q., & Wells, C. K. (2014). Untapped richness in Erik H. Erikson's rootstock. *The Gerontologist, 54*(1), 40–50. doi:10.1093/geront/gnt123 PMID:24136991

Kübler-Ross, E., & Kessler, D. (2005). *On grief and grieving*. New York, NY: Scribner.

Marchant, J. (1646). *Rationale evangelizantium, sive doctrina et veritas evangelica, a sacerdotibus, pastoribus, concionatoribus, pectori appendenda, plebique per anni circulum è Cathedris proponenda, exponenda: Duobus tomis comprehensa*. Paris, France: Apud Michaelem Soly.

Meinecke, L. D. (2017). *Neglected by assessment: Industry versus inferiority in the competition for scarce kidneys* (Doctoral dissertation). Available from ProQuest Dissertations and Theses database. (ProQuest No. 10689852)

Meinecke, L. D. (2018a). A theory of cognitive idolatry, or the human struggle to preserve psychological certainty in a biological species - chapter 1 (Research proposal). doi:10.13140/RG.2.2.16527.43684

Meinecke, L. D. (2018b). A theory of cognitive idolatry, or the human struggle to preserve psychological certainty in a biological species - chapter 2 (Research proposal). doi:10.13140/RG.2.2.31089.53601

Meyer, M., & Bergel, K. (2012). *Reverence for life: The ethics of Albert Schweitzer for the Twenty-First Century*. Syracuse, NY: Syracuse University Press.

Panksepp, J. (2007). Can PLAY diminish ADHD and facilitate the construction of the social brain? *Journal of the Canadian Academy of Child and Adolescent Psychiatry*, *16*(2), 57–66. Retrieved from https://www.ncbi.nlm.nih.gov/ PMID:18392153

Rochat, P. (2003). Five levels of self-awareness as they unfold early in life. *Consciousness and Cognition*, *12*(4), 717–731. doi:10.1016/S1053-8100(03)00081-3 PMID:14656513

Rogers, C. R. (1961). *On becoming a person*. Boston, MA: Houghton Mifflin.

Ryckman, R. M., Thornton, B., & Butler, J. C. (1994). Personality correlates of the hypercompetitive attitude scale: Validity tests of Horney's theory of neurosis. *Journal of Personality Assessment*, *62*(1), 84–94. doi:10.120715327752jpa6201_8 PMID:8138889

Saint-Exupéry, A. (1943). *Le petit prince*. New York: Harcourt, Brace & World.

Search, P. W. (1901). An ideal school, or looking forward. In W. T. Harris (Ed.), *International Education Series* (Vol. 52, pp. 1–357). New York, NY: D. Appleton and Company.

Sharot, T., Korn, C. W., & Dolan, R. J. (2011). How unrealistic optimism is maintained in the face of reality. *Nature Neuroscience*, *14*(11), 1475–1479. doi:10.1038/nn.2949 PMID:21983684

Spinoza, B. (2012). *The ethics* (R. H. M. Elwes, Trans.). Start Publishing.

Stone, S. D., Crooks, V. A., & Owen, M. (2013). Going through the back door: Chronically ill academics' experiences as 'unexpected workers'. *Social Theory & Health*, *11*(2), 151–174. doi:10.1057th.2013.1

Tolstoy, L. (2019). *How much land does a man need?* Retrieved from http://www.online-literature.com/tolstoy/2738/

Vazquez, C. (2013). Positive psychology: Introduction to the special issue. *Terapia Psicológica*, *31*(1), 5–9. doi:10.4067/S0718-48082013000100001

ENDNOTES

1 When the newest Diagnostic and Statistical Manual of Mental Disorders came out a few years ago, there was much ado about the addition of a new moratorium on normative grieving.

2 The imago is the fully developed stage of an insect, or an unconscious idealized mental image of someone (Imago, 2020).

Section 2

The Paradox of Environmental Protection

This introduction to the second book section defines the purpose of this second part. The four chapters included in this section include "The Healing and Grounding Potential of Biophilia," "National Park Theory," "Colonialism Disguised as Protection," and "At the Crossroads: Sovereignty or Faith?" This introduction simply outlines the themes that the second set of four chapters will cover, serves as one of the dividers of the material into three logical chunks, and helps outline how they fit into the organization of the book's three objectives.

The theme of this second section is to introduce the reader to the many ways an untrammeled wilderness can be of equal benefit to this bipedal species, without the constant need for its subjugation and domestication. Too often the benefits of collaboration deny the minority shareholder anything but the consequences. Nature has been too long the silent partner of a domineering partnership, and the failure to share in both the lasting virtues and external goods equally afforded yet unequally distributed has left the planet with little to live on save its penury (Annas & Wang, 2008). Disguised as mercy, expansionism continues in the use of the environment as a playing piece in global affairs. Yet humans also have a very good side, a side that sets aside large parcels of land for the enjoyment of its beloved people and the preservation of the natural beauty of its territory (Muir & Highland, 2001). Could such a conservation, generalized to the whole Earth in the form of a vast park, become the solution to the climate crisis? Perhaps the card that has not been played by those who need to predict the future to save it is Chance after all, not better predictions. Perhaps a little faith in the planet's ability to manage its own affairs with the best interests of its non-human people in mind has been the solution all along.

REFERENCES

Annas, J., & Wang, H. L. (2008). Aristotle on virtue and happiness. *Ancient Philosophy (Garland Publishing)*, *37*(2), 347–359. doi:10.5840/ancientphil201737228

Muir, J., & Highland, C. (2001). *Meditations of John Muir: Nature's temple*. Berkeley, CA: Wilderness Press.

Chapter 5
The Healing and Grounding Potential of Biophilia

ABSTRACT

Love of nature involves going out into nature and building a beneficial relationship with other living things. Often this results in a person suddenly feeling grounded, centered, focused, and assured. Forest therapy is one way to find this sense of belonging and solace, but isolation and loneliness can still pervade even the most sacred of spaces. The authors posit that what all living things need is to feel important to one another. Sadly, the civilized world does not often grant that regard. Still, a silent companion calls from the wild to come home; this chapter is about several ways to answer that call and get back both one's regard for nature and one's own self-regard. Biophilia is a proven way through the pathless woods of depression, a trail unmarked by the signposts of civilization—where wildflowers with high hopes of loveliness still grow.

INTRODUCTION

Grounded vs. Centered

Upon first glance, the phrase, *being grounded in Nature* appears to be redundant and in dire need of editing; however, Dr. Diana Raab (2017) differentiated between being *grounded* versus being *centered*, and further proposed the possibility of individuals being both grounded *and* centered. A

DOI: 10.4018/978-1-7998-4408-2.ch005

sense of calm appears to be associated with the two terms, but they are not synonymous although they do often tend to accompany each other. Being *grounded* is defined as having the ability to remain completely aware and conscious within the present time, or it is defined as practicing *mindfulness* where "what-ifs" are rarely entertained. The author, Michael Daniels (2005), describes *groundedness* as the sense of being completely embodied and whole within oneself that is accompanied by feelings of harmony, clarity, and 'rightness.' A person who is grounded has total control of his or her mental and emotional state where he or she is not easily impacted by others, their ideas, comments, etc. which allows inconsequential mishaps to roll off his or her back like water rolls off of a duck, so there is much less of an inclination to become reactive or overwhelmed by something or someone. In the event that there is a situation where a typically grounded individual does feel "off balance," there are several strategies that can be used to re-establish the grounded *status quo*:

- Focusing one's attention on breathing; meditating while walking
- Playing instrumental music without the distraction of words
- Practicing reflective listening
- Sipping a hot beverage, such as tea, being aware of its warmth and taste
- Engaging in multi-sensory awareness, which can become an entry in a journal to reinforce the experience
- Giving non-human family members one's full and undivided attention to show them love and appreciation

In contrast, being *centered* is defined as having a reference point which can be a place to which an individual can return when life becomes especially challenging and stressful. The center to which an individual *must* return can be represented by the medicine wheel that is often referenced in Native American culture where each of the four directions representing the physical world is joined in the center of the wheel. The north direction represents the mind while the south direction represents the heart; the east direction represents the spirit, and the west direction represents the body, so to maintain a sense of balance or well-being in body, mind, and spirit, all four directions must be in sync. Focused breathing is a strategy that enables an individual to return or his or her center just as it facilitates returning to groundedness (Raab, 2017), but in addition, Roshi (2017) also suggested three of the tenets related to the practice of Zen which are: "Not knowing, bearing witness, and taking action" (p. 3). The first of these three tenets, *not knowing*, involves letting go especially during turbulent circumstances or situations; the second tenet, *bearing witness*, involves acknowledging or being mindful of the earth's joy and suffering; and the third tenet, *taking action*, involves being one with the intention so that the best action will be taken for any given situation. According to Buddhist meditation, bearing witness (tenet two) is spontaneous and concerns being aware of thoughts and feelings as they arise, but allowing them to pass by in cloudlike fashion. Similarly, taking action (tenet three) is also unpredictable but should always be caring and considerate for oneself as well as for anybody else who is involved. Gaining a basic comprehension of being grounded and centered when coupled with some simple strategies can result in an overall sense of well-being in body, mind, and spirit.

FOREST THERAPY

In March 2019, Hannah McQuilkan, a qualified Naturopath, Medical Herbalist, and Forest Therapy Guide with The Association of Nature and Forest Therapy emailed this author a copy of her article entitled *Forest Therapy and Three Prisons in Three Countries*. According to Hannah, Forest Therapy consists of a series of optional invitations designed to connect practitioners more completely with Nature by reducing the pace and simply (mindfully) *being* in Nature. She describes one of the *Pleasures of Presence Guided Journeys* as an invitation to create a signature fragrance from the surrounding plants, and then sharing them with the rest of the group members. Typically, the Forest Therapy experience is concluded with a tea ceremony of found plants, such as fresh chamomile and lemon balm, from the garden which helps to calm the body and mind while mitigating the urge to ruminate about the past as well as to worry about the future. Hannah suggests that this interest and involvement in Forest Therapy is a kind of Renaissance that hails back to a time when the Earth was inhabited globally by indigenous groups who comprehended the inextricable link between humans and Nature that was absolutely vital to the life and well-being of all. Many individuals subscribe to the belief that they already spend time in Nature working, exercising, socializing, etc. in it, but what they do not acknowledge is the fact that their agenda is hindering their experience to generate a more intimate one-on-one relationship with Nature. Based upon the volume of research that supports the existence of the fundamental connection between humans and Nature, it simultaneously appears to suggest that depriving individuals of this vital connection, no matter what the context (home, school, work, prison, etc.), is a violation of human rights. Multiple studies have indicated that individuals who regularly engage with Nature experience the following benefits: Faster recovery from illness; a reduction in anxiety and/or depression; extended lifespan; and increased performance across all areas.

LONELINESS AND ISOLATION: THE DISTRESSING BIOPSYCHOSOCIAL EFFECTS OF WITHDRAWING FROM NATURE

Typically, when considering the categories and levels of risk or danger encountered within a society or a community, the focus is on the frequency and classification of criminal statistics, such as violent crime (homicide, assault, battery, etc.) versus non-violent crime (breaking and entering, robbery, etc.). However, a national survey conducted by Cigna in 2018, determined that levels of loneliness have climbed to a record high with almost half of the 20,000 adults in the U. S. reporting that they always or occasionally felt alone. In addition, forty-percent of these survey participants reported that they always or occasionally felt isolated and felt as though their relationships were not meaningful (Novotney, 2019). According to Holt-Lunstad Smith and Layton, (2010), loneliness exacerbates physical and mental health risks to the level of smoking 15 cigarettes per day or to demonstrating an alcohol use disorder. Dr. Holt-Lunstad also determined that loneliness and social isolation were twice as detrimental to physical and mental health as obesity (Holt-Lunstad, Smith, Baker, Harris, & Stephenson, 2015).

Several campaigns and coalitions have been organized globally in an effort to mitigate an individual's perceived level of social isolation and loneliness by raising awareness through advocacy and interventions. The question is: Has social isolation and loneliness actually increased or have they existed in the human condition all along so that only the recognition and willingness to discuss them has increased? Although historical data pertaining to social isolation and loneliness are difficult to find, some research suggests that they are on the increase as evidenced by recent U. S. census information which stated that in excess of one quarter of the U.S. population lives alone, which is the highest level ever recorded. Additionally, greater than half of that population is single or unmarried while marriage rates as well as the number of children in each household have decreased since the preceding census. Social connections that are usually provided by membership in religious, organizational, or volunteering opportunities have also decreased, so whether social isolation and loneliness have increased is not truly the crux of the issue, but rather the fact that a significant number of members of the population has been negatively impacted by them *is* the problem. Even from an evolutionary perspective, an individual's well-being and ultimate survival has been determined to be based upon his or her ability to establish and to maintain social connections (Holt-Lunstad et al., 2015); however, this author posits that these social connections must extend beyond only human interactions and encompass all of Nature (O'Grady, 2016).

Clinical psychologist Ami Rokach purported that while advocates, scientists, and public policy-makers may be poised to assist individuals, as well as society in general, in combatting their feelings of social isolation and loneliness, they are facing an escalating public health crisis with the prospect of an increasing aging population of Baby Boomers and post-Boomer generations. In addition, loneliness is a facet of the human condition where at some time or other, everyone has to cope with this type of *reactive loneliness* during periods of transition, such as moving to a new location or during periods of loss due to death or divorce (Novotney, 2019). However, more severe issues are likely to arise when reactive loneliness becomes *chronic* or downright tortuous, which can develop in circumstances where individuals do not have the resources to emerge and satisfy their emotional, mental, or financial needs; or they do not have a social support system that can help them achieve these goals (Hawkley & Capitanio, 2015). In 2018, Pew Research Center surveyed 6,000 adults who correlated their frequent bouts of loneliness to their dissatisfaction with family, community, and social life. Rokach also commented that loneliness can exist even when individuals are in crowds or surrounded by others, and loneliness is not synonymous with solitude or desired isolation but is measured by their dissatisfaction with their perceived levels of connectedness or social isolation (Novotney, 2019). The adverse effects of perceived loneliness and isolation, according to Kassandra Hawkley, in addition to depression include inferior quality of sleep, diminished executive function, impaired immunity, accelerated decline in cognition, poor cardio-vascular function across the lifespan. A recent 2019 study that was weighted for such factors as marital status, frequency of religious service attendance, club meetings and/or social group participation, as well as the number of close relatives and friends suggested that social isolation increased the risk of premature death for all races from every cause (Alcaraz et al., 2019).

Finding solutions and developing interventions to combat the adverse effects of loneliness and isolation which have been well-documented in this previous research tends to be a challenging undertaking since there is no single underlying cause for loneliness and isolation. At different stages across the lifespan, individuals are likely to experience these feelings for different reasons which would then require a different intervention or solution that has been tailored to fit the situation. The list of potential remedies to address feelings of loneliness and isolation include external and internal strategies (respectively) for improving social skills, enhancing social support networks, increasing opportunities for social interactions, in addition to *cognitive behavioral therapy* (CBT), which is designed to identify and address negative perceptions regarding self-worth and others' opinions. The engagement of individuals in community and social groups such as choirs, book clubs, or church groups depending on their interests, willingness to learn, and experience can increase positive mental health effects while decreasing feelings of loneliness and isolation. Globally, *cohousing* seems to be increasing in popularity, as cohousing communities and/or mixed-age residences have been constructed to purposely bring older and younger generations together in neighborhoods, apartment buildings, or single-family homes to share dining areas, recreational spaces, etc. Neighborhoods organize events, form clubs and organizations, and make provisions for childcare and eldercare while simultaneously mitigating feelings of loneliness and isolation through this inter-generational housing strategy (Novotney, 2019).

GREEN SPACES AND MENTAL HEALTH

But what if, as these authors posit, despite the multiple underlying causes that exist for loneliness and isolation, as well as their related maladies, there is a universal solution for reducing and/or eliminating these levels of distress which can have potentially disastrous consequences in the more severe cases? And what if that universal solution could be as simple as taking a walk outdoors or spending time in a favorite *sit spot*, with or without meditation, to allow each individual to *be* cognitively present *in the present* which, in the vernacular of the 1960s, was described as "communing with Nature." A longitudinal research study published in 2019 by Engemann et al. (2019) supported that premise by affirming that the integration of natural environments into urban planning and development provides a positive approach to the improvement of mental health and the reduction of the global rise of psychiatric disorders. Engemann et al. (2019) investigated the prospective relationship between green space and mental health in the Danish population of over 900,000 residents through the use of high-resolution satellite data which calculated the normalized vegetation difference index surrounding each individual's place of residence to the distance of 210 x 210 m square from the time of birth to age 10. Their results indicated that high levels of green space are correlated with lower risks of a diverse variety of psychiatric disorders later in life (a negative statistical correlation). Conversely, the risk for higher levels of a diverse variety of psychiatric disorders later in life increases with access or exposure to lower levels of green space and was 55% greater across the various psychological disorders even after adjusting for SES (socioeconomic) factors, urbanization, and parents' age and history of mental illness. The research results also indicated that there was a stronger association to lower risks of psychiatric disorders with cumulative green space presence across the span of childhood versus the exposure or access to one single year of cumulative green space presence. Engemann et al. (2019) hypothesized that during the earlier childhood years, the pathways involved in passive exposure to noise reduction and/or pollution removal may be important, in contrast to the pathways involved in actions such as exercise and/or socialization as children function more independently. In addition, parents' actions and schools' facilities will encourage children's visits to parks and other types of green spaces. Future research studies might be designed to capture data relating to biodiversity: The quality of vegetation, blue space, and the presence of animals within the green space under investigation. The loss or reduction in the frequency of opportunities for human interaction with Nature not only increases the potential for developing mental health disorders later in life, but it also diminishes an individual's appreciation of natural environments. Therefore, utilizing green spaces within the design and planning of urban settings are likely to provide mental health benefits, while simultaneously serving to protect biodiversity and natural environments.

It appears as though current advancements in the field of complementary and alternative medicine (CAM) have finally caught up with the past history of human evolution when approximately six to seven million years ago, when the species of human beings had existed less than 0.01% in modern surroundings, in contrast to the other 99.99% of the time living in/with Nature. With this statistic, it is little wonder that many humans are drawn to the point at which human physiological and psychological functions began and were supported by Nature (Hansen, Jones, & Tocchini, (2017). Kellert and Wilson's *Biophilia Hypothesis* (1993) and O'Grady's *The Theory of Biophilia* (2016) support Shinrin-Yoku and Nature Therapy which are anchored in the concept that humans possess a *biophilia gene* which makes them innately hard-wired in their attraction to Nature, as well as its significance to human development. On the spiritual and psychological levels, human beings are intuitively aware of the relaxation, soothing, and awe-inspiring effects created by being in or seeing forests, plants, flowers, green spaces in urban areas, parks, and even natural wooden materials.

SEEDS FOR THOUGHT: BIOPHILIC APPLICATIONS

Gardening versus *planting with purpose* involves more than just sowing seeds for providing nutritious food for human and non-human consumption; *planting with intent* involves total sensory awareness in connecting with Nature through the smell of the freshly tilled soil, the sight of the emerging green shoots, the tactile pleasure of coming into contact with the Earth, and the physiological, psychological, and spiritual rewards of nurturing life that should never be taken for granted. In addition, maintaining such an intimate interaction with Nature teaches lessons in recycling of natural materials, such as constructing homemade pots for starting seeds from empty paper towel and/or toilet paper tubes. These are especially suitable for seeds which prefer a long running root system, such as beans and peas. Plants offer several ecological benefits, such as providing shade, shelter for wildlife, and food. In addition, plants offer healing benefits for the body, mind, and spirit. Jessi Bloom (2019) notes that human beings have evolved using plants for medicinal purposes, wellness, and nutrition; in addition, she notes that modern pharmaceuticals are often formulated to mimic natural herbs or are formulated using natural herbs in conjunction with synthesized compounds. Bloom purports that there is a positive correlation between the ability of the human body to utilize the healing properties of herbs when they are closest to the herbs' natural state. So, dedicating some time to learning about the healing properties of herbs and adaptogens, as well as their growing requirements, can result in the construction of an herbal garden suited to the available space (backyard plot, sunny apartment window, rooftop garden, etc.), and is limited only by an individual's creative imagination. Specific herbs were initially classified as *adaptogens* in 1947 by N.V. Lazarev, a Soviet scientist, and in 2007 were defined by herbalist, David Winston and researcher Steven Maimes as a group of herbs which assists the body in adapting to stress while supporting regular metabolic processes, and restoring balance to the system while causing few or no side effects in most of the population. Today adaptogens are preferred by traditional and alternative healing modalities globally because their efficacy increases the longer they are used, often with greater and broader benefits. Although adaptogens do not target any one particular issue, they support a person's collective body system(s) by adapting to the symptoms and disruptions created by the disease/disorder whether they are emotional, physiological, and/or environmental (Combs, 2019).

Bloom (2019) suggests utilizing a type of anthropomorphizing approach when choosing which plants to include in any designated space. By approaching them as people with genetically pre-determined needs often aids in identifying their ideal growing conditions and characteristics. Note that all plants have originated from somewhere, so by knowing which conditions will allow them to flourish will be mutually beneficial for the grower and the "growee" (plant). Rather than trying to re-invent the wheel, careful consideration of the existing environment can be especially helpful in choosing the plants most suitable for the current conditions, such as areas with only limited access to sunlight where less than six hours of direct sunlight is classified as being a *shade garden*. For areas that are challenging to maintain, irrigate, etc., an uncultivated or *wild medicinal garden* consisting of plants that demonstrate a natural competition with one another eliminates the necessity of more formal cultivation. For individuals who reside in cold climates or in urban apartments but have a sunny windowsill, an *indoor medicine garden* becomes a viable option when adhering to a few simple rules: choose plants with similar soil and watering requirements; choose deep pot saucers for hydration and to mitigate spillage; fertilize during the plants' growing season; and repot into larger containers every one to two years to allow for root expansion. Allowing plants to become established and flourish for up to three years provides an excellent guideline for knowing when to harvest from them. However, lovingly tending those plants with gratitude, and harvesting them with positive intention goes far in encouraging them to provide their strongest medicine.

HOMELY PRODUCE

According to statistics released by the U. S. Department of Agriculture, by all rights, hunger should not exist in the U.S., but even more disturbing numbers suggest that Americans waste up to 40% of the food they grow annually. The

U. S. Department of Agriculture reports that in 2016, in excess of 41 million Americans were categorized as being *food insecure*, or lacking consistent access to a sufficient amount of affordable and nutritious food, and of those people were 13 million children. Globally, the U.S. produces sufficient food to feed 10 billion individuals; however, almost one billion people go without due to inadequate production and distribution strategies which results in waste at every level of food production from the farms, to the supermarkets, to the restaurants, and to the homes. Each of these levels maintains its own criteria for rejecting and/or wasting food, but the underlying reasons for this waste is based upon aesthetics, where a lot of produce is discarded; not because it is inedible but because it is ugly and does not adhere to the rigid cultural standards for beauty. Really? The standard for food production has become about as distorted as the cultural or societal norms for beautiful people. This assessment has little to do with the healthful value or contribution of the food or of the individual. Imperfections in the world of produce can range from knobby-kneed carrots, to bruised bananas, and to scarred strawberries. Any type of bruise, discoloration, softness, or hole are all offenses that cannot be tolerated, and that lands a perfectly healthful and nutritious fruit or vegetable in the garbage. Americans have become so obsessed with the aesthetic beauty of food that it has developed into a phenomenon and national pastime of photographing food for Instagram (MacLean, 2019).

A certain degree of sanity has prevailed in a growing tide against the pursuit of food perfection in the form of many proactive organizations within the U. S. that are working to divert "ugly food" from the waste streams, which have adopted the old name, *gleaning*. This charitable act of feeding the poor dates back to the old testament of the Bible, and is also mentioned in the Quran, and ancient rabbinic literature. Up until the latter part of the 18th century, landless European citizens were afforded legal protection to glean, but in the latter half of the 20th century, it became common for farmers, restaurants and supermarkets to prohibit this practice because they feared liability and prosecution in the event that an individual would become sick from this discarded food. In 1996, the Bill Emerson Good Samaritan Act served to reduce the liability that potential food donors could endure due to intentional and/or gross negligence which reinitiated the gleaning culture in such pursuits as *dumpster diving* (MacLean, 2019).

Gleaning organizations function directly with everyone who is involved in the food supply chain, such as orchards and supermarkets, and arranges to pick up food that has passed its sell-by date but that is still good. Local food pantries then distribute it to individuals who live in virtual *food deserts*, so while the efforts of the gleaning organizations and local food pantries are indeed commendable, they barely make a dent in the endless stream of wasted food. A more effective strategy would be to divert the needless waste at the level of homes by meeting the stream at the source of farms and farmers who avoid bringing "seconds" or ugly produce to market since there is a reduced likelihood of it selling. Although these "seconds" are typically composted or fed to the farm animals, efficiency would certainly be increased if this food were consumed by humans, for whom it was originally intended. Consumers have the option of approaching sellers at their local farmer's markets to request bulk orders of seconds at a discounted cost which benefits both the farmers and the consumer. The resurgence of interest in ugly produce has also resulted in a resurgence of interest in canned and fermented food processing, where the generational inheritance of food preservation knowledge is once again providing pleasure to new generations of home cooks (MacLean, 2019). Personally, this author enjoys fruits and vegetables that elicit a bit of humor with their less-than perfect appearance, and feels that each one should be prepared with a sense of gratitude toward Nature.

WHAT'S THE BUZZ?

According to the articles in Industry News and Bring on the Bees published in the April/May 2019 issue of *Mother Earth Living* magazine, the Earth's global food network is dependent upon the survival of plant and animal species, not the least of which are bees. The mission of *Bee City USA*, which is a nonprofit initiative of the Xerces Society, is dedicated to the construction of sustainable pollinator habitats within urban environments. Communities are encouraged to collaborate with Bee City USA, and to become affiliates with the goal of raising public awareness regarding threats facing pollinators, enhancing their habitats to allow them to flourish, and showing them easy ways in which they can help the initiative. Community participation in the Bee Campus USA programs results in the improvement of the community's overall environment, as well as the local food production, and it benefits small businesses and local plant nurseries. Translating the importance of pollinators into statistical representation indicates that there are in excess of 4,000 species of native bees in North America which are responsible for more than 20 billion (with a "b") dollars of food crops annually. Many of those indispensable pollinator species are experiencing declining populations due to exposure to pesticides, loss of habitat, and/or disease, such as bee colony collapse; and while continuing to support European honeybees is important, assisting our native North American bees is even more imperative. Even urban gardeners who maintain small patio or balcony gardens in containers can create sustainable habitats for these pollinators out of readily available materials: galvanized trash cans, buckets, lumber for a nesting box, a piece of burlap, bamboo gardening stakes, and/or wire and twine, which all are a small investment that pays major dividends when forming a partnership with pollinators.

NATURE'S PATHWAYS TO INNER PEACE

Some individuals center or ground themselves by running or hiking while other individuals practice yoga, but walking a labyrinth provides yet another form of moving meditation spiraling in upon itself and then back out again to its perimeter in the guise of a meandering path that helps to clarify the mind. While the terms, *labyrinth* and *maze* are often used interchangeably, they are not synonymous since the former clarifies in contrast to the latter, which is deliberately constructed to confound the individual. Mazes are described as a design that contains multiple entrances and exits, dead ends, and confusing paths in contrast to labyrinths which are designed as a single pathway leading to its center and back out to its edge or perimeter, so there is only one way to walk with no decisions needed regarding which way(s) to turn. The premise underlying a labyrinth is to allow the walkers to focus solely on their inner thoughts for the purposes of centering, grounding, prayer, and/or to identify their yearnings and aspirations. Labyrinths date back four millennia, and their patterns appear in ancient petroglyphs, drawings, woven baskets, and hedges in various cultures and religions, such as England, Greece, Russia, Indian, and Christianity (Caldwell, 2019).

Labyrinths have experienced a renewal in interest in recent years, and are thought to provide healing, joy, and peace to individuals in their times of grief and confusion. Although complex inlaid tile designs, such as the one found in the Chartres Cathedral, which continues to attract pilgrims to this day (please see: https://www.atlasobscura.com/places/labyrinth-chartres-cathedral) and may be categorized as works of art. Backyard labyrinths can be created out of almost anything including plants, a pattern mowed in the grass, or groundcover (Caldwell, 2019). With the continued support of such non-profit organizations as The Labyrinth Society, which hosts a website at https://labyrinthsociety.org, plans for the design and construction of personal labyrinths and are readily accessible via the Internet. Sue Mosher, who conducts guided labyrinth tours in throughout Washington, offers the following simple step-by-step tips as a starting point for walking a labyrinth:

First: Identify a specific reason or purpose for your journey, or simply remain open to the experience.

Second: At the threshold, or the beginning, express your intention, take a deep breath, and relax. Remove your shoes to be in closer contact with the path if desired.

Third: While walking the labyrinth, be aware of the alternating rhythm of the straight sections, turns, and movements toward and away from the center. Release any concerns or expectations that you may be harboring.

Fourth: Avoid rushing, and linger in the center of the labyrinth to receive whatever message or information may be waiting for you there. Carry it gently with you upon your return journey (Colbert, 2014).

Mosher, a trained Veriditas labyrinth facilitator, explains that an individual can seek multiple things simultaneously when walking a labyrinth, such as centering oneself by letting the body know he/she is about ready to walk the path. Just as there is no one specifically correct material for constructing a labyrinth, there is no one specifically correct way to walk one. An individual's left brain knows the directions to walk in and out without getting lost, and so is able to relax a bit more while the right brain or the intuitive side can find the answers to the questions being sought by the individual. Walking releases the information, so by the time the center of the labyrinth has been reached, the individual may have a clear answer to the question or concern, or he/she might simply feel a momentary release to the tension. Although the left brain is partially occupied with navigating the correct path, yet not having to actively seek it, the majority of the processing ability will be focused to intuit solutions. Walking the labyrinth is not typically part of an individual's daily routine, therefore, that experience can be utilized in locating the solutions that already exist within his/her brain, which allows the individual to rejoin the world and apply those findings and/or solutions (Caldwell, 2019).

In the early 1990s when the AIDS crisis had reached an all-time high, the Rev. Lauren Artress in San Francisco discovered a labyrinth that had been created by Dr. Jean Houston, and found it to be a very powerful tool for solace that people could perform together without having to talk. She saw firsthand how walking a labyrinth provided a sense of calm and cohesion to the residents of a city that had been torn apart. In 1995 Artress founded a nonprofit organization in California, *Veriditas*, which translates into "the greening power of life" and is dedicated to 'inspiring personal and planetary change through the labyrinth experience' (Caldwell, 2019, p. 60). She oversees the installation of labyrinths all over the world as well as training labyrinth facilitators worldwide, hosting events, and travel that are designed to encourage individuals to seek a deeper comprehension of the potential for healing and transformation that labyrinths can provide. Veriditas and The Labyrinth Society have collaborated in creating a global labyrinth index which can be found via the following link: www.LabyrinthLocator.com. It contains archival data relating to more than 5,700 labyrinths in over 80 countries which are in outdoor gardens, at public centers, not to mention private individuals' backyards (Caldwell, 2019).

SACRED GEOMETRY

The Rev. Artress attributes a portion of the healing properties of labyrinths to the shape or arrangement of their paths. The width should not exceed the measure of the shoulders of a human being. This narrowness permits walkers to assume their natural walking cadence, and to simply follow the spiral along, while their minds process the queries that they have brought into the labyrinth. Globally, cultures attribute symbolism to spiraling shapes which suggest movement, growth, expansion, and change around a central point surrounded by a singular path. Classical types of labyrinths possess seven circuits which is also a number that is reminiscent of the seven visible planets, chakras, musical notes/tones, and colors; however, by medieval times, labyrinths were often created with 11 circuits surrounding a 12th, central circle, with four quadrants constructed by the 180-degree turns in the path. In the Gothic labyrinths, the numerological significance extends to the number of turns built into the path with the six-petaled flower frequently located at their center and the *lunations*, or small semicircular decorations, inscribed into the labyrinth's outermost path. In contrast, modern labyrinths tend to incorporate significant numbers from many traditions, and often experiment with two paths at the beginning which ultimately merge at the center. No matter which labyrinth design offers the most appeal, each one offers a unique opportunity for individuals to reconnect with himself/herself, Nature, and the surrounding community. They can be constructed to fit any space or budget, and can be as temporary or permanent as desired, but attention should be paid to the number of people who are anticipated to be walking the labyrinth at one time (Caldwell, 2019).

CONCLUSION

Using Nature to find healing means finding a connection between oneself and the natural world—feeling grounded, centered, focused, and assured. Forest therapy is one way that a person can find healing out in Nature. But isolation and loneliness can still pervade even those sacred spaces. What is needed is to be around others, be they human or non-human, who make a person feel as important as Nature itself. That can help transform the sense of being *different* to a feeling of being *needed*; that same transformation often allows a person to help others—be they human or non-human—do the same. Planning ahead with green spaces is another way that biophilia can comfort and heal when civilization steals away an erstwhile natural assurance in everything around you—that you are not alone. Planting with purpose can convert an ordinary garden into a long-term companion, and such practices may have unexpected benefits for non-human species such as bees. Living labyrinths may provide much needed answers for short-term goals and navigating them is a pleasant activity. Research suggests there is almost a sacred geometry between the search for natural assurances and a lovely public labyrinth where people can find a cadence in which unseen symmetry seems to conform to both the needs of one's tangible body and one's intangible soul.

REFERENCES

Alcaraz, K. I., Eddens, K. S., Blase, J. L., Diver, W. R., Patel, A. V., Teras, L. R., ... Gapstur, S. M. (2019). Social isolation and mortality in US black and white men and women. *American Journal of Epidemiology, 188*(1), 102–109. doi:10.1093/aje/kwy231 PMID:30325407

Bloom, J. (2019, March/April). Landscaping with medicinal plants. *Mother Earth Living*. Retrieved from https://www.motherearthliving.com/

Bring on the Bees. (2019, February 15). *Mother Earth Living*. Retrieved from https://www.motherearthliving.com/

Caldwell, N. (2019, March/April). Seek serenity in a backyard labyrinth. *Mother Earth Living*. Retrieved from https://www.motherearthliving.com/

Colbert, J. (2014). *Are DC's labyrinths a path to inner peace?* Retrieved on June 28, 2019, from https://www.washingtonian.com/2014/04/30/are-dcs-labyrinths-a-path-to-inner-peace/

Combs, D. (2019, March/April). Flourish with adaptogens. *Mother Earth Living,* Retrieved from https://www.motherearthliving.com/

Daniels, M. (2005). Shadow, self, spirit: essays in transpersonal psychology. Charlottesville, VA: Imprint-Academic.com.

Engemann, K., Pedersen, C. B., Arge, L., Tsirogiannis, C., Mortensen, P. B., & Svenning, J. C. (2019). Residential green space in childhood is associated with lower risk of psychiatric disorders from adolescence into adulthood. *Proceedings of the National Academy of Sciences of the United States of America, 116*(11), 5188–5193. doi:10.1073/pnas.1807504116 PMID:30804178

Hansen, M. M., Jones, R., & Tocchini, K. (2017). Shrinrin-yoku (forest bathing) and nature therapy: A state-of-the-art review. *International Journal of Environmental Research and Public Health, 14*(8), 851. doi:10.3390/ijerph14080851 PMID:28788101

Hawkley, L. C., & Capitanio, J. (2015). Perceived social isolation, evolutionary fitness and health outcomes: A lifespan approach. *Philosophical Transactions of the Royal Society B, 370*(1669). doi:10.1098/rstb.2014.0114

Holt-Lunstad, J., Smith, T. B., Baker, M., Harris, T., & Stephenson, D. (2015). Loneliness and social isolation as risk factors for mortality: A meta-analytic review. *Perspectives on Psychological Science, 10*(2), 227–237. doi:10.1177/1745691614568352 PMID:25910392

Holt-Lunstad, J., Smith, T. B., & Layton, J. B. (2010). Social relationships and mortality risk: A meta-analytic review. *PLoS Medicine, 7*(7), e1000316. doi:10.1371/journal.pmed.1000316 PMID:20668659

Kellert, S. R., & Wilson, E. O. (1993). *The biophilia hypothesis*. Washington, DC: Island Press.

MacLean, K. (2019, March/April). Preserve farmers market seconds. *Mother Earth Living*. Retrieved from https://www.motherearthliving.com/

McQuilkan, H. (2019, March 11). *Forest therapy and three prisons in three countries*. Retrieved from https://treemystic.com/forest-therapy-and-three-prisons-in-three-different-countries/

Novotney, A. (2019). Social isolation: It could kill you. *Monitor on Psychology*, *50*(5), 32. https://www.apa.org/monitor/

O'Grady, M. A. (2016). *The theory of biophilia* (Reg. No. TXu 2-032-922). United States Copyright Office. Retrieved from https://www.copyright.gov/

Raab, D. (2017). Are you grounded? Centered? Or both? *Psychology Today*. Retrieved on May 3, 2019, from https://www.psychologytoday.com/us/blog/the-empowerment-diary/201705/are-you-grounded-centered-or-both

Roshi, W. E. N. (2017, Summer). Hold to the Zen advice for when things blow up around you. *The Buddhist Review,* 36-39. Retrieved from https://tricycle.org/magazine/

U. S. Department of Agriculture. (2016). *Agricultural statistics 2016*. Retrieved from https://www.nass.usda.gov/Publications/Ag_Statistics/2016/Complete%20Ag%20Stats%202016.pdf

Chapter 6
National Park Theory

ABSTRACT

Chapter 6 delves into the topic of regard. In order to begin the process of reconciliation with the planet, mankind must first ask what is wrong with his broken relationship with the Earth. If man's first date with nature was one of wonder and awe, could it be that man now takes his beloved nature for granted? Perhaps his silent partner would like a say in how she is regarded, and especially in how she is treated. If so, what can man do to get back to that first love and shower the weeping Earth with affection? It turns out that the national park system may point the way back to conjugal bliss.

INTRODUCTION

A human being is a part of the whole called by us "Universe," a part limited in time and space. He experiences himself, his thoughts and feeling as something separated from the rest, a kind of optical delusion of his consciousness. This delusion is a kind of prison for us, restricting us to our personal desires and to affection for a few persons nearest to us. Our task must be to free ourselves from this prison by widening our circle of compassion to embrace all living creatures and the whole of nature in its beauty" – Albert Einstein

A lot of focus, these days, is about challenging human thoughts. Maybe it's time to challenge human feelings instead? After all, a thought is more like a feeling that just won't go away. So, why have everyone's feelings gone away? Why don't humans feel pity for their dying planet? They want a restraining

DOI: 10.4018/978-1-7998-4408-2.ch006

order, then a divorce. They're fed up with Nature. It's like their first love has died, and humans can't seem to cry. Amy Grant used to sing about it, with memorable tunes like *Where Do You Hide your Heart?* and *Stay for a While* (Grant, Kirkpatrick, & Smith, 1986; Grant & Smith, 1984). Where *do* humans hide their hearts when it comes to their planet? Why can't everyone just tarry here a while, in this lovely garden, and watch the sunrise together? As the Reverend Billy Graham used to say at his big crusades in Oakland, "Your friends will wait, your buses will wait." The Earth seems to be saying, "Don't give up on me. These moments are precious. Every moment matters. Stay for a while, and sip a cup of coffee with me, your green-eyed bride—before you hurry off to wherever it is you need to go. You and I can watch the Sun come up together, and just chat for a while."

In this chapter, the reader will explore why humans have lost their feelings for their beloved world. The Earth was humanity's first love, but now its feelings have died. Imagine nobody is giving up. Imagine the planet and humanity are going to see a marriage counselor and find out what went wrong, and see about fixing this relationship. The human species is going to find its awe again (Keltner & Haidt, 2003), and get back that very first love (Gould, 2011). But how, the reader may ask? How can anybody fix this?

The authors have a theory. It's called *National Park Theory*. In a nutshell, it posits that when two people are still new to one another, their time together is like a gift. Each feels like a visitor to the other, and each is like a National Park the other has come to visit. Both are in profound awe of their surroundings—and Time seems to stand still. But all too soon, when the two get to know each other too well, when their once amazing surroundings become too familiar, one of them begins to feel like he *deserves* the other. Something dies inside—a thing called gratitude, and a thing called awe. Too often, he begins to see the one he loves as his property, instead of being so grateful for her he can barely speak. She leaves. When time at home is no longer rare or special, it is taken for granted; perhaps it is time to remember how special home used to be? This author is fortunate to live right next door to a National Park; it reminds him never to take Her for granted, and to be grateful for a home, every day.

FIRST IMPRESSIONS AND FINAL IMPRESSIONS

First of all, what is missing in this relationship? *Regard* is missing. Mankind used to love Nature; now he acts like she is in his way. Love begins with regard, and regard comes in basically two forms: temporary impressions, and lasting impressions. The former is known as amnesia (Meinecke, 2020). The latter is known as prejudice (Biernat & Danaher, 2013; Plous, 2002). Temporary regard is what one has every moment, so long as one can't remember what happened a moment ago. This can be a good thing, if one can't remember what was pleasant (because it will be pleasant again, whether one can remember or not). It can also be a very bad thing, if one can't remember what hurt (because it will hurt again, whether one can remember or not; Holmes, James, Kilford, & Deeprose, 2010).

Lasting regard, on the other hand, is like a first impression that just won't go away. Post-traumatic stress disorder (PTSD) is like that. Losing your wife to cancer is like that (you can only lose her once; but you lose her again and again each time you wake up and she isn't there). That happened to this author. Like the phantom limb effect, if the two parted under pleasant circumstances, their lasting regard is either one of endearment or of profound sorrow (Meinecke, 2018). But if all they did was cause one another pain, a sudden severing of their ties preserves a lasting sense of phantom pain that won't go away (Ramachandran, Rogers-Ramachandran, & Cobb, 1995).

Okay, the first question is what's wrong with this relationship? The next question is why is the human attitude toward human works so different from the human attitude toward Nature/Creation? What is missing from Mankind's attitude toward the planet he lives on? By definition, an attitude is "an enduring pattern of evaluative responses toward a person, object, or issue" (Colman, 2009, p. 64). Isn't this the definition of prejudice? After all, as the saying goes, "There but for the grace of God go we" (Pedersen & Thomas, 2013). Although a bad attitude is not what anyone would wish upon themselves, a bad attitude *is* a lasting conviction about someone who wishes they would forgive them. In a more basic sense, it is an orientation toward something; one is attracted to it or repulsed by it, whether or not that thing is here. An attitude is like a lasting impression then, but with an attitude—an even more resistant opinion which is very hard to undo. Beliefs are attitudes too, although beliefs generally refer to mental concepts rather than physical things.

How does a first impression become a belief? Maybe it's simply the result of trying to *create* a favorable impression that will always feel the same? Unfortunately, first impressions of anything vary widely, so that a lasting impression is like a first impression that cannot vary, and attitudes and beliefs do seem very invariant. One way to leave a lasting good impression is to part ways when neither wishes to part. Conversely, a good way to leave a lasting bad impression is to part ways when neither can stand the other, and one or both can't wait to be apart. Now the last thing one remembers is the way it felt the last time together. This is the foundation of the phantom limb theory—to amputate a limb when it doesn't hurt—not when it hurts so badly you can't wait to part with it. That way, if that parting is done gently and amicably, the wound won't keep hurting as though the severed limb were still there crying out "Why are you leaving me?" Kahneman (2011), a recipient of the Nobel Prize, also posited this and showed some evidence of its truth. In cognitive psychology, these principles are referred to as *primacy* (first impressions), *frequency* (averaged impressions), and *recency* (last impressions), and they do seem to influence people whether they want them to or not (Goldstein, 2015; Sternberg & Sternberg, 2016).

THE NEUROSCIENCE OF REGARD

But what is an impression? Does everything people see leave an indelible image in their brain? The jury is out on that, but some say "yes" and others say "no." The ones that say no believe it's more of a mental proposition than a mental image (Thomas, 2019). Whatever the case, every moment leaves a first impression. One just can't remember them all. Sadly, the human mind exhibits a fundamental bias, in that it is inclined to notice what it's used to (reinforcers), more than what it isn't used to—so it becomes very hard to forget how painful an object is, even after it becomes pleasant. The mind under-attends non-matching stimuli, and over-attends matching stimuli (Goldstein, 2015). Can lasting impressions be trusted then? The answer is "no," according to some neuroscientists.

This author attended a class on neuronal dynamics hosted by Lausanne, and it was greatly suggested by the lecturer that humans never really see an object so much as a river of different signals that get averaged into a solid object. He used the Sydney Opera House as an example, and showed how there isn't just one opera house, really, so much as lots and lots of opera houses that get reduced down to become one "average" opera house (Gerstner, Kistler, Naud, & Paninski, 2014). There is collaborating evidence from psychology, as well, that suggests humans heap perceptions of what is always unique outside the mind, into concepts that appear to be similar inside. It lets humans make sense of far too many stimuli. It is a well-known principle which psychologists of the Gestalt era pointed out, like grouping things as similar that are obviously not exactly the same, or mindlessly closing an unclosed figure, or compulsively completing the other half of a tiger behind a tree even if one can't be sure there is another half (Chapman, 2020).

It seems feasible that the need for similitude itself underpins the very concept of enduring bias, and that nothing is ever similar to itself twice if it was significant once. This was once proposed by prominent scientific minds such as Weber and Fechner, and called the *just noticeable difference* or JND (Hergenhahn & Henley, 2014). The JND of anything in psychology is called a first impression in the vernacular. Some researchers have actually extracted an eye from a representative member of a species and used it as a telescope to imagine a Copernican model of the inner workings behind that eye (Carter, Aldridge, Page, & Parker, 2009). After all, the perception of a group of similars is just an objective interpretation of subjective sensory stimuli constrained by shared physiological and conceptual limits.

FEEDFORWARD ATTITUDES AND BALLISTIC BELIEF SYSTEMS

So, the idea of perception is already an opinion, and an opinion is merely a statistic aggregated from rigid observations among conspecifics over time, who naturally have trouble seeing things that vary significantly from their common *line of best fit* (Witte & Witte, 2004). If perception is an opinion though, there must be the actual thing there is an opinion about—the thing itself—or (more famously) the *Ding an Sich* as Immanuel Kant put it. The object of an opinion should have a say in its definition. Nature (the thing itself) should have a say in Mankind's definition of Nature (its attitude toward that thing). Science frequently calls these two ways of deciding what something "is" either a *feedforward process* (unilateral agreement) or a *feedback loop* (bilateral agreement).

A feedforward process is typically when one thing perceives another thing (observation) without asking the other thing to confirm its opinion. One might call this a ballistic opinion or belief system, because once it is underway, nothing can alter it (other than to get out of its way). Prejudice is a lot like this. The Earth cannot contest the human species' demeaning attitude toward it. Humans don't really let the things they observe have much of a say in what they think they are[1]. That's kind of unfair when you think about it. Humans often put what they observe into a category they happen to have—without asking *what they observe* how it feels about being put in that category. It isn't always possible to ask things to confirm whether a guess is right; the investigator, like the librarian in Searle's Chinese Room, may not speak the language or pass a Turing test certifying his claims to know how Nature feels (Cole, 2020). They may not even have a language, let alone need one—any more than a rose that's classified as a "rose" needs to know it's a rose to behave like one.

In any case, in neuroscience, this is often called *feedforward*: "the modification or control of a process using its anticipated results or effects" ("Feedforward," 2003). Some call this *motor out*. What all this means, is that humans don't really care what it might actually *be* so much as what humans *think* it is based on what it *does*. With a feedforward perspective, rather than wait for feedback, whatever humans think or do is a purely outgoing act, and what happens as a result of that observer-centric thought or act humans don't really care much about. Oddly, things do have a tendency to conform to human expectations if humans wait long enough (because humans don't believe in the alternatives). By example, the unrepeatable neuronal ripples that arise from viewing the same object over a period of time (neuroscientists call this the *subthreshold regime*), eventually reduce down to imaginarily enduring patterns in that ever-fluctuating ripple (Gerstner et al., 2014). These become semi-finite patterns (*schema*) that can be recalled and manipulated as internal world objects for simulating outcomes safely (this topic is called *knowledge representation* in cognitive psychology; Brewer, 2000). It's hard to recall a thing before it becomes a reliable thing to recall. To see a pattern, you must already believe there is one. This suggests that belief often precedes the evidence, neglecting anything which does not align with the expectation.

DIVINE HIDDENNESS

Why would humans be right in front of something and keep looking for it? As the authors stated above, when humans have objectified everything, they "cannot see the forest for the trees" as the saying goes. If humans already "know" what humans are looking for, they won't be able to see it unless it corresponds exactly to what humans "know." Similarly, if humans say it doesn't exist, they won't believe they see it even if it's standing right in front of them. By reverse, if humans hold that a thing is absolutely true, no volume of evidence disproving it will get in. Attitudes are like that. Prejudice is like that (Biernat & Danaher, 2013; Plous, 2002). If Mankind's feelings are missing, then, perhaps it's because Mankind's thoughts won't let them feel them? Frequently it seems that human thoughts are—jealous—of the real things they let humans think about, making them feel "guilty" for counting using their visible fingers when they "should" be counting invisible sheep in their heads.

The authors of this book are pursuing research into hidden object games for just this reason—the proposition that it is actually *harder* to find something when you know exactly what you are looking for. Please see: http://lmeinecke.com/biophilia/. In brief, say the hidden object game asks you to find a bat. In your head you will more than likely have a mental image of a baseball bat (if you were raised in a culture where baseball is a big deal, rather than spelunking). The authors have found that, when people have an idea of what to look for, if the object does not exactly match that idea, they won't see it. So if the game designer placed a *Fledermaus* there (literally "flying mouse"—a flying mammal of the order *Chiroptera*), it will take a person longer to spot it than if it were a Lil' Slugger (if they spot it at all). Gobs of mental certainty often detract from spotting what those with just a smidgen of uncertainty easily see. The authors' research on this topic should be coming out soon!

That said, very often humans remain near familiar things, so that their confirmation bias[2] remains content with its opinions (Pettigrew, 1979). Chances are, there are no patterns hidden in nature per se, so much as the human need to see them is so great that when humans most need them, they are always there. Even the famous dual-slit experiment suggests that something at the base of Nature is aware of the need for confirmation, and becomes whatever it is humans most needed Nature to see or do (e.g. a particle or a wave—both, if necessary; Avant, 1965). One can guess this is fundamentally true, this proclivity to think magical patterns are everywhere in Nature, because humans see things that aren't there all the time (they are called *pareidolia* and *apophenia*—the detection of patterns that simply aren't there). Please see the authors' video at: http://lmeinecke.com/videos/3-Pareidolia.mp4.

Humans do not actually see objects repeatedly; their minds fool them into thinking there are constant "objects" that vary over time, rather than variable stimuli they very much wish would remain constant (Gerstner et al., 2014). This is probably due to something called *mortality salience* or *terror management theory*, which is the anxiety of losing what has come to feel dear to you (even if you will eventually lose everything, since all humans perish eventually; Harmon-Jones et al., 1997). It is also helpful for something called *object permanence*, a memory of things no longer evident, so that one can use them to perform inner calculations concerning external things one does not otherwise understand or control (Baillargeon, Spelke, &

Wasserman, 1985). Please see the authors' video at: http://lmeinecke.com/videos/2-ChangeBlindness.mp4. Even human beings are inconstant, but see themselves as more constant than they really are because the thought that they perish every moment is too terrifying to bear (Gilbert, Sapp, & Tauber, 2012; Harmon-Jones et al., 1997). Yet, humans needn't be afraid of dying. Moser (2002) put it this way:

Many people are perplexed, even troubled, by the fact that God (if such there be) has not made His existence sufficiently clear. This fact – the fact of divine hiddenness – is a source of existential concern for many people. That is, it raises problems about their very existence, particularly its value and purpose. The fact of divine hiddenness is also, according to some people, a source of good evidence against the existence of God. That is, it allegedly poses a cognitive problem for theism, in the form of evidence challenging the assumption that God exists. (Howard-Snyder, Moser, & Daniel, 2002, p. 1)

Now, one can frame the problem another way. Let the word "Nature" (Creation) be substituted for the word "God." It would read like this: "Many people are perplexed, even troubled, by the fact that Nature (if such there be) has not made its existence sufficiently clear" (paraphrase of Moser, 2002, p. 1). Maybe, in order to matter, one must be perceived. Maybe, to be perceived (thus to truly matter and be needed), one must first be imagined (Berkeley, 1979). Once imagined, you have an idea of what to look for.

TO BE PERCEIVED, PERCHANCE TO MATTER

The struggle for existence, in large part, is to sort of grapple with regard. If you can garner somebody's regard (anybody's regard), somebody will make sure you don't perish. Berkeley famously called it *the need to be perceived*, which is a lot like the need to matter more than anything else, to somebody or something (Berkeley, 1979). A famous passage from scripture called this state of being needed *the pearl of great price*:

Again, the kingdom of heaven is like a merchant seeking beautiful pearls, who, when he had found one pearl of great price, went and sold all that he had and bought it. (Matt. 13:45-46, NKJV)

This old parable still seems sound; if you've found what you're looking for, why would you keep looking? Is there more than one of the best thing of all? To be regarded like that must be an awesome feeling. That would definitely make life seem worth living. Just imagine if someone regarded each and every person that way. As a famous positive psychologist used to say, "Other people matter, and we are all the other people" (Peterson, 2008).

So, if "to be perceived" was a big thing in Bishop Berkeley's day, "to matter" must be a big thing today. What are some synonyms in psychology? Is there a scale to see where one stands that isn't associated with anything other than one's existence? There do seem to be a few terms for "mattering" in psychology, although it is not certain whether there is a validated scale wherein humanity's regard for someone's simple existence is measured. Two of these terms, however, are *valence* and *attribution.*

Valence (aka hedonic tone) is psychology's term for the "goodness" or "badness" of a thing, which unfortunately means lots of things won't matter in a good way (if they matter at all). Valence is more associated with choices and judgments than existential appreciation, and unnatural drives such as goal attainment and norm violations. As Nature probably does not set goals or worry about violating social norms, valence would not be a very suitable measure for reverence for Life.

Tropism would be a better choice, in keeping with the natural sciences (the relative attractiveness or aversiveness one thing seems to hold for another, without assessing goodness, badness, correctness, or deficit). Tropism is more about what is needed to grow and thrive, rather than an emotional judgment about what should or shouldn't be attractive (Freud treats this in a study of taboo). It is curious that humans do not attribute poor choices to non-human species; it is also curious that humans do not think non-human species matter. But Life does seem to be attracted to itself, and away from Death, whereas humans are very much attracted to Death and market it in films, games, outdoor sports, and real life (Fromm, 2013).

Attribution goes beyond valence, in a way, by suggesting this is not just how humans feel about the thing, rather this is what the thing "is" (and it "makes" humans feel that way). In this manner, humans can disown their feelings and blame them on the object. (Again, this is very Freudian, as in his description of *narcism*, now termed narcissism). But it isn't an inanimate thing's fault humans feel this way; humans simply project the feelings they are ashamed of into the things they think they should be ashamed of. The authors covered this more fully in chapter 3, *How Taxonomy Steals Reverence*, and both Sigmund and Anna Freud discuss mental behaviors quite similar to these in their work on ego defense mechanisms and taboos (Freud, 1966; Freud, 1920; Freud & Brill, 1913).

Disowning one's feelings by attributing them to some fault in something else, is what many call prejudice (Biernat & Danaher, 2013; Fiske & North, 2014; Plous, 2002). It is the heaping of many differences into a single belief held by one side of a two-sided need for regard.

ON A COMMENSAL PERCEPTION

Ideally, though, perception is not decided by one half of two things. In order to matter, there is a mutual regard held between all things, rather than by one thing for everything else. With the environment, if humans feel that the environment does not care how humans feel about it, it is easy to see why humans regard ourselves as fairer.

Can Nature indicate how it feels about people? Maybe it already does? Curiously, another famous neuroscientist—the late Dr. Jaak Panksepp—wondered if maybe people aren't listening correctly. Maybe people just can't hear the joy in Nature using equipment that has evolved to pick out only human voices? He found that, during play, rats were figuring out one another's boundaries, and both humans and rats can tell just how much is "too much," because they will either giggle or cry out (Panksepp, 2007). He used a special listening device to hear the laughter of rats. A lot of sciences use this feedback approach as well; scientists make a guess, but they don't stop there. They also "ask" the percept if they might be right (scientists don't just keep measuring their own guesses and ignore the feedback). Based on the feedback science gets from the object, the scientist revises his or her perception, and guesses again. Scientists keep doing that until their guesses and the feedback match. In fact, Niels Bohr and Albert Einstein had a famous argument about mutual perception versus simple observation. They were talking about the moon. Einstein is reported to have asked Bohr, "Do you think the moon is not there when we aren't looking at it?" To this Bohr replied, "Can you prove to me the opposite?" (Pais, 1979, p. 907). Maybe the moon cares whether people miss it or not when it isn't there. Maybe Nature is not all that inanimate?

The reader may note that, in non-classical physics (which is what amazing scientists like to explore in their spare time), it just might matter whether humans *want* to see something or not. Maybe the moon is only there when it *needs* to be there, and somehow that's the only time it *is* there? Maybe if a person could just look out of the corner of his or her eye (when the moon isn't ready), it would vanish? Certainly, people act a whole lot differently when they know somebody is watching than when they think nobody is watching (Piazza, Bering, & Ingram, 2011; Sheldrake, 2005). Maybe non-human things aren't a whole lot different than humans are? After all, everything is made of the same stuff, isn't it? Humans are just arranged differently. Maybe the quantum of a thing is just enough of the thing needed to be perceived, just as the need to see it is just enough for Nature to matter enough to exist—and thus counterbalance those two potentials? After all, Bishop Berkeley suggested that the most fundamental need of any extant thing is *to be perceived* (Berkeley, 1979). Maybe extant things *make* themselves visible to the eye (like stereotype threat in a minority group) so that the eye will at least see them a little (rather than not at all)? That would certainly help explain the dual-slit experiment (sometimes the eye needs to see a particle; sometimes the eye needs to see a wave). In any case, it's a terrible thing to be invisible when you know you're there but nobody else seems to notice you. Familiar things often regress to unnoticeability (Avant, 1965), hasn't the reader noticed? So, if so, that would mean that the essential nature of perception is that it should always be a mutual affair… and observation really is unfair to the stuff humans observe because observation alone doesn't give them a chance to see if there is any feedback—and thus adapt their attitude toward them. Maybe Nature isn't there unless a person needs to see Her?

THE BIOLOGICAL TEMPLE: BODY AS EARTH

Here, there is a need to bring up a very key concept. The authors call the concept *the biological temple* (in the current theory of biophilia). What do the authors mean by that? Well, it is the most curious thing that the way most humans see the physical body they inhabit, is very similar to how they see the physical planet they inhabit. For example, if they think of their body as a resource or tool, they probably think of their world as a resource or tool. If they think of their body as a precious temple, though, they probably think of the world they live on as a precious temple too. Similarly, in a functional sense, if they view their body as something to be tamed and controlled to help the mind achieve its internal goals, they probably see (and treat) their environment the same way. Both of these are like vessels; humans seem to use their bodies to get around, and their body seems to use the world to get around. James Gibson called this view of material things "affordances" (Gibson, 1977).

The thing is, if a person learns to see his or her body (or planet) as a tool, resource, or stepping-stone to bigger and better things (*an affordance*), he or she will eventually see both their body, neighbors, and their world as stepping stones too (this is called an instrumental or utilitarian worldview in psychology). Perceiving things this way is usually called *prejudice*, because it leads to a condescending view of those things. It leads to a sort of egotistically facilitating worldview, rather than to see other species more holistically and adoringly (as equals in a full and unranked, ecological picture). Many Western beliefs originated from an Aristotelian worldview. This was later iconicized during the Roman Empire and Middle Ages when there was an unusual need to justify the great "distinction" between things like higher species (gods and angels), the slightly lower but still god-like species (humans), and the much lower and ungodly species humans needed to exploit on their way to becoming angels (Gilhus, 2006). These odd taxonomical habits were needed to justify widespread inequality among overpopulated humans, the horrifying transformation of animals into food using sacrificial rituals, and of course, a vast new need for uncompensated labor to build enormous empires. It is not surprising that Aristotle came up with the *Scala Naturae* (Ladder of Life) to explain Mankind's favored rung on the ladder of Life—with an unmistakable disregard for both human flesh and animal flesh, and a strange, sudden overemphasis on the overweight brain of Man (Holzman, 2018; Paleos.com, 2019; Ramachandran, 2012). See Figure 1, The Great Chain of Being, in chapter 3. Curiously, Aristotle belittled the older idea of *hedonia* (seek pleasure and avoid pain, because life is brief), and seemed to dismiss the preciousness of these brief moments, with the promise of unreachable mental rewards—such as lasting virtue and external goods[3] instead of transitory, physical joy (Annas & Wang, 2008).

So, if human beings are taught to believe that their invisible mind is superior to their visible body, they are very likely to believe that their invisible ideals are also superior to their visible world and its hapless animals. Some Canadian psychologists have actually demonstrated this is so, and shown that children can unlearn this terrible prejudice, with benefits to their distorted attitudes toward other human groups too (Costello & Hodson, 2014). If humans think they are somehow "more amazing" than everything else in Creation, they will probably think it their "duty" to displace the imperfect, natural world to make more room for their amazingness. But if all men of conscience should look at their own bodies with humility and awe, it follows those men might view their natural environment with a similar humility and a similar awe. And if humans could just see these brief bodies as living temples of unimaginable worth—not as their immortal proprietors, nor as their redeeming priests, but as timid, adoring worshippers *in this brief temple*—it follows that those humans might extend the same reasoning to that endless biological Temple which is this sapphire orb everyone adores. Would such a perception change the way humans treat their environment? The authors think it will.

Many grow up thinking their body is more like an animal that needs to be tamed and controlled—and act more ashamed of what they physically are than what they psychologically expect from that animal vessel. Those same individuals gather into unusual interest groups, and oddly maintain that the entire Earth they live on is like an unruly beast as well. They seem angry at pity of any kind, and mandate that it be subdued and forced to be subservient to the ambitions of a selfish species. This callous perspective is apparent in the myopic treatment of (and soon the imminent death of) this "savage" (yet less selfish) biological planet—while that "civilized" and (and much more egocentric) species continues to fight over what's left of its dying world. So many avid political platforms act as though, no matter what they do to it, neither they nor this world will ever perish, that Mankind's resources will never run out, and that everybody can be fabulously wealthy if they work hard enough. Many see their own body the way a dog trainer sees a pet that needs to be on a leash, or as a form of inalienable mental property that must not be emancipated (no matter how much it pleads for its freedom). They may become frustrated that their body/pet does not do exactly as they say. And when they get old, and the body cannot keep up anymore, they compel others to act as eerie simulacra for their overwrought body which no longer blindly obeys their commands (Baudrillard & Glaser, 2014). They do not seem to notice how they regard their own flesh with such scorn (nor how they disregard the lives of other humans and animals in the path of their mental objectives).

In contrast, some would never see their physical body as a heritable affordance nor as a disobedient pet, but as a precious gift no matter how briefly they have it. Some treasure the time they have with their day-old infant, though its brief candle goes out that night. And some see their body more like a sort of living temple. In fact, they see everybody's body that way, and respect living things just because they live. Somehow, this attitude translates to regard of the living planet itself as a kind of Natural Temple and engenders a reverence for Life of every kind (that includes both their own flesh and the flesh of every living thing). Instead of regarding the planet as a resource, they see it as an endless majesty full of vibrant creatures and forces. They seem less likely to abuse either their own body (or their world) the way some do in their struggle for fame and glory, or the pitiable struggle to cope in a world with few lawful ways to survive save substances and habits that end up as unlawful regrets in the end. Instead, they adore both self and world; they marvel at organisms and cells; they marvel at organs and ecosystems and climate; they cherish and would never sell what many would rather sell to the highest bidder (the National Parks). Some have even begun to sell what to most are national treasures—natural monuments set aside because, initially, no one would have dreamed of selling them, nor any commercial rights to them.

NATIONAL PARK THEORY

The authors have another theory that fits alongside the biological temple. It's called *National Park Theory*. It's more than just a theory, it's an objective. The objective is to make the whole world a national park. Why? The world is dying, but it doesn't have to. Why not rezone the entire world as a park, but bit by bit—not all at once. The basis is simply that *social momentum* is a powerful influence (see also *psychological momentum* and *behavioral momentum*), so that once this momentum gets going, land use may reach a tipping point beyond which Nature has enough of a range to survive (Nevin, Mandell, & Atak, 1983). Humans may even carry this banner further to its logical limit—and designate it all as a park, to be enjoyed by its heirs forever.

As to the specifics, muncipalities often designate land use using ordinances and zoning codes, such as R for residential, C for commercial, and P for parks and similar public lands. Even within a National Park system, land use may be subdivided using ideas such as conservation, use, and special use (Mulyana, Moeliono, Minnigh, Indriatmoko, & Limberg, 2010). The authors propose that Humankind consider the global rezoning of the planet Earth to "P" (and for the reader familiar with Stephen Hawking's work, the authors propose the Somewhere of P as opposed to the Elsewhere of P; Hawking, 1998, p. 27). The authors further propose that the whole Earth be designated for conservation and reverence only (no use of public lands for anything but wonder and gratitude).

Every day Humankind can tell this world how grateful they are by offering Her some of Her land back- and do it for joy, not because people have to. These gifts will forever after become the indefeasible property of a National Park shared by everything on Earth, the citizen and the non-citizen, the indebted and the debt-free, the human and the non-human animal. These species will become its visitors, caregivers, and adorers, and living glory—instead of its bickering proprietors and vendors.

This lovely planet is dying. Scientists like E.O. Wilson (2016) have argued that humans can shift away from the tipping point of loss of all life on this world, if humans simply give the planet back to Nature. Nature knows how to fix itself. The most intelligent species does not seem to know much about fixing this problem, nor understand the idea of doing a good that is *so good* Life of every kind survives (no matter whether it benefits the human species or not). As mentioned above, post-Aristotelian humans only understand deferred joy, and scorn a universal *hedonia*. Instead of seeking joy right now, modern humans pursue "virtues" and "external goods" (wealth, friends, and political power) to preserve human fame rather than for a planet to plant those names upon. Nature as a whole does not seem to think or act this way—our thoughts and ways are not Nature's thoughts or ways (Isa: 55: 8-9). Life is not concerned with its fame nor either with its infamy. All things struggle to exist for joy right now, while they live, rather than later when others enjoy what they bled and died for. As Housman (2020) put it:

Eyes the shady night has shut
Cannot see the record cut,
And silence sounds no worse than cheers
After earth has stopped the ears. (stanza 4)

This reluctance to put the Present[4] first puts humans at a huge disadvantage, because it precludes solutions frequently discarded as "unsustainable" or "economically infeasible." This theory is an option that makes saving the planet feasible. Humankind's reward is the planet's survival.

Perhaps it is time to digress for a moment. There is a phenomenon in which, when humans stand to lose a little, they become bitter about it (that's another theory by the authors—*The Hurricane Survivor Effect*, which simply posits that those who lose a little get upset, and those who lose everything but each other become grateful). Even though E.O. Wilson's plan is a good one (to give back half the Earth), folks are unlikely to give back even a little of their land. Would you go first if you thought no one else would follow, and you would be laughed at?

As the Hurricane Survivor Effect theory posits, when humans lose a little, or must give up a little, humans become very resentful and stingy. They feel cheated. But for some reason, when humans lose everything, they are often grateful just to be alive—and look around and treasure the fact that their families are okay (even if their homes and property are a disaster). This author lived through a set of tornadoes in Tennessee, and observed how human kindness responded to near death experiences. When humans incur damage due to a natural disaster, they rue even that little damage. But when they lose everything but one another to a natural disaster, they rue nothing. The come together in quiet humility, and rejoice in rebuilding their community. Could this fact be applied to the current climate crisis?

So, if this species responded (right now) as though this species had lost it all already (as though the end of the world had already happened), and looked around and suddenly realized everything will be alright, maybe this species would be glad to give its holdings back to Nature out of gratitude for being alive? Scientists have noted that, when humans let Nature decide what's best for Nature, Nature begins to heal itself in ways humans hadn't thought of. Things go back to the way they were before humans plowed everything under and paved away Nature's means to recover. Imagine that this species could look forward to waking up every morning in the biggest National Park ever reserved for living things? This author and his family lived in a little home, nestled in a village full of forests, critters, and lakes, all next to a National Park in Arkansas. It's hard not to be inspired whenever you look up from your studies. The primary author of this book inspired this co-author, to consider this idea of changing how people feel about their planet—from one of domination to one of awe. When someone loves something, really loves something (no matter what happens), that someone suddenly sees just how amazing that something is. The authors call this "awed by Nature." The world is so worth saving.

National Park Theory posits that, if humans could view themselves as visitors blessed to briefly visit this planet, not proprietors of the planet it as though it were their property, nor as though they were its sole and rightful heirs, they might begin to change. This species might experience a change of heart and a change of attitude toward the Earth, and enjoy an abundant awe and gratitude for each moment they get to spend with this sacred land and its inhabitants. If humans could but view this world as a huge national park they have been permitted to enjoy, they would never mistreat Her and spoil the chance for their children to adore Her. It is like the way this author learned to view his sweet wife Marcy (who died prematurely of cancer, sigh)—to see her with a reverence he had not felt before. When a man does not mistake what he's lucky to have at all for a right, or a possession, or an unlosable expectation, he begins to see her with awe, and becomes grateful for each hour he has with her.

There are some side benefits too. If humans would give back their land to a stochastic authority, a good many land use issues (and their catalyzation of human conflict factors) would vanish as well. Without such boundaries around the world, is any creature a stranger, trespasser, or nuisance animal in the land it briefly occupies? Need any distrustful interest watch or defend its borders against its neighbors, if it hasn't any borders? Parts of New England and parts of the South have a lovely tradition. They do not build fences between one another's yards, nor worry about where that man's yard ends, and this man's yard begins. They often honor one another more than the legal boundaries that divide them. Imagine if every neighbor of every person enjoyed the same right to that universal honor? Perhaps then human beings of every kindred and in every region could spend more time honoring each other rather than honoring their borders. This author's wife was not from here, was never born with the rights this author enjoyed; yet he learned to value her more than all he had enjoyed. Imagine if every human saw every other human being this way? When Life is first, somehow who has the right to it doesn't matter.

Again, to digress a bit, it seems fitting to extend this idea of what matters most in Life. Why do humans treasure their words more than the real things their words refer to? Why do they worship the circumscription of their territory and revere its name, and look upon the human and non-human inhabitants of that territory as inferior and expendable so long as the named territory survives? Perhaps it is a deliberate accident, if one may coin an oxymoron for this. If humans did not circumscribe the land, humans would not honor the borders more than what lies either side of those borders. This does not honor anybody. These enduring concepts mislead many into missing their own brief worth. American poets and explorers have long adored the once vast frontier of unconquered creatures, un-parceled land, in words as elegant as John Muir's (Muir & Highland, 2001). Rather than a reverence for what humans have done, there is another kind of reverence for things humans did not do—yet it is a gorgeous and thriving work, despite the fact humans did nothing to create it or deserve it. That too is how a citizen feels who did not have to become naturalized to enjoy his freedom, and met someone who did.

CONCLUSION

In this chapter, two primary areas of concern have been explored. One, does it matter how humans regard the human-planet relationship? Yes. It matters a lot. And two, if it does matter, is there anything humans can do to improve this relationship? Yes. There is a lot humans can do. Could it be that by improving the way the human species views its world, the approaching crisis might be slowed or even reversed? Absolutely.

Does it matter how a husband sees his wife? Yes, it does. A man who feels overjoyed to marry the woman of his dreams does not forget that first awe, but awakens every morning next to her like a visitor in paradise. A poem about a flea by John Donne drives home the irony of human choice (Donne, 2020). Even a flea enjoys the right to what humans deny one another. So, since in this chapter it has been established that it matters, is there a way to improve the way humans see their beloved planet too? Yes, there is. John Muir and other famous naturalists noticed that the human environment is more like a bride than a resource, and fell madly in love with her. This chapter took a good hard look at this from a modern psychological perspective. When a man or woman feels like a visitor to his or her significant other, he or she tends to treat that other with awe. But when it feels like what they love is theirs by right, they seem to forget how much they once loved one another. One day, this arrogant species may even perceive its beloved world the same way it had once humbly perceived the woman of its dreams… and return (at last) to that first tender love. As the song goes, "It's not too late to get it back again" (Gould, 2011, 1:52).

REFERENCES

Annas, J., & Wang, H. L. (2008). Aristotle on virtue and happiness. *Ancient Philosophy (Garland Publishing), 37*(2), 347–359. doi:10.5840/ancientphil201737228

Avant, L. L. (1965). Vision in the Ganzfeld. *Psychological Bulletin, 64*(4), 246–258. doi:10.1037/h0022208 PMID:5318040

Baillargeon, R., Spelke, E. S., & Wasserman, S. (1985). Object permanence in five-month-old infants. *Cognition, 20*(3), 191–208. doi:10.1016/0010-0277(85)90008-3 PMID:4064606

Baudrillard, J., & Glaser, S. F. (2014). *Simulacra and simulation*. Ann Arbor: The University of Michigan Press.

Berkeley, G. (1979). *Three dialogues between Hylas and Philonous* [Kindle ed.]. Indianoplis, IN: Hackett Publishing.

Biernat, M., & Danaher, K. (2013). Prejudice. In I. B. Weiner (Ed.), Handbook of psychology (Vol. 5, pp. 341–367). Hoboken, NJ: Wiley.

Brewer, W. F. (2000). Bartlett's concept of the schema and its impact on theories of knowledge representation in contemporary cognitive psychology. In A. Saito (Ed.), *Bartlett, culture and cognition* (pp. 69–89). New York, NY: Psychology Press.

Carter, R., Aldridge, S., Page, M., & Parker, S. (2009). *The human brain book*. New York, NY: DK Publishing.

Chapman, C. (2020). *Exploring the Gestalt principles of design*. Retrieved from https://www.toptal.com/designers/ui/gestalt-principles-of-design

Cole, D. (2020). The Chinese room argument. In E. N. Zalta (Ed.), *The Stanford Encyclopedia of Philosophy*. Retrieved from https://plato.stanford.edu/archives/spr2020/

Colman, A. M. (2009). *Oxford dictionary of psychology* (3rd ed.). New York, NY: Oxford University Press.

Costello, K., & Hodson, G. (2014). Explaining dehumanization among children: The interspecies model of prejudice. *British Journal of Social Psychology*, *53*(1), 175–197. doi:10.1111/bjso.12016 PMID:23083220

Donne, J. (2020). *The flea*. Retrieved from https://www.poetryfoundation.org/poems/46467/the-flea

Feedforward. (2003). In *Oxford Dictionary of English*. Retrieved from https://www.lexico.com/en/definition/feedforward

Fiske, S. T., & North, M. S. (2014). Measures of stereotyping and prejudice: Barometers of bias. In G. J. Boyle, D. H. Saklofske, & G. Matthews (Eds.), *Measures of Personality and Social Psychological Constructs* (pp. 684–718). doi:10.1016/B978-0-12-386915-9.00027-9

Freud, A. (1966). *The ego and the mechanisms of defence*. New York: International Universities Press.

Freud, S. (1920). *A general introduction to psychoanalysis*. New York, NY: Horace Liveright. doi:10.1037/10667-000

Freud, S., & Brill, A. A. (1913). *Totem and taboo*. New York, NY: Dover Publications.

Fromm, E. (2013). *Escape from freedom*. New York, NY: Open Road Media.

Gerstner, W., Kistler, W. M., Naud, R., & Paninski, L. (2014). *Neuronal dynamics: From single neurons to networks and models of cognition*. Cambridge, UK: Cambridge University Press. doi:10.1017/CBO9781107447615

Gibson, J. J. (1977). The theory of affordances. In R. E. Shaw & J. Bransford (Eds.), *Perceiving, Acting, and Knowing*. Hillsdale, NJ: Lawrence Erlbaum Associates.

Gilbert, S. F., Sapp, J., & Tauber, A. I. (2012). A symbiotic view of life: We have never been individuals. *The Quarterly Review of Biology*, *87*(4), 325–341. doi:10.1086/668166 PMID:23397797

Gilhus, I. S. (2006). *Animals, gods, and humans: Changing attitudes to animals in Greek, Roman, and early Christian thought*. London, UK: Routledge. doi:10.4324/9780203964798

Goldstein, B. E. (2015). *Cognitive psychology: Connecting mind, research, and everyday experience* (4th ed.). Belmont, CA: Wadsworth Publishing.

Gould, K. (2011). *First love. On Roots of Love* [CD]. Nashville, TN: Fine Linen Music.

Grant, A., Kirkpatrick, W., & Smith, M. W. (1986). *Stay for a while. On The Collection* [CD]. Nashville, TN: Myrrh Records.

Grant, A., & Smith, M. W. (1984). *Where do you hide your heart? On Straight Ahead* [CD]. Los Angeles, CA: Warner Chappell Music, Inc.

Harmon-Jones, E., Simon, L., Greenberg, J., Pyszczynski, T., Solomon, S., & McGregor, H. (1997). Terror management theory and self-esteem: Evidence that increased self-esteem reduced mortality salience effects. *Journal of Personality and Social Psychology*, *72*(1), 24–36. doi:10.1037/0022-3514.72.1.24 PMID:9008372

Hawking, S. W. (1998). *A brief history of time*. New York, NY: Bantam.

Hergenhahn, B. R., & Henley, T. B. (2014). *Introduction to the history of psychology* (7th ed.). Belmont, CA: Cengage Learning.

Holmes, E. A., James, E. L., Kilford, E. J., & Deeprose, C. (2010). Key steps in developing a cognitive vaccine against traumatic flashbacks: Visuospatial Tetris versus verbal Pub Quiz. *PLoS One*, *5*(11), 1–9. doi:10.1371/journal.pone.0013706 PMID:21085661

Holzman, L. (2018). *The overweight brain: How our obsession with knowing keeps us from getting smart enough*. New York, NY: CreateSpace Independent Publishing.

Housman, A. E. (2020). *To an athlete dying young*. Retrieved from https://www.poetryfoundation.org/poems/46452/to-an-athlete-dying-young

Howard-Snyder, D., Moser, P., & Daniel, H. S. (Eds.). (2002). *Divine hiddenness: New essays*. Cambridge University Press.

Kahneman, D. (2011). Thinking fast and thinking slow [Kindle ed.]. New York, NY: Farrar, Straus, and Giroux.

Keltner, D., & Haidt, J. (2003). Approaching awe: A moral, aesthetic, and spiritual emotion. *Cognition and Emotion*, *17*, 297–314. doi:10.1080/02699930302297 PMID:29715721

Meinecke, L. D. (2018, April 14). The uncanny fear of loss, part 1: Losing the unthinkable. *Psychology Today*. https://www.psychologytoday.com/us/blog/theory-and-praxis/201804/the-uncanny-fear-loss-part-1

Meinecke, L. D. (2020, March 7). *Nature-nurture and language development*. Retrieved from psyarxiv.com/4u3kq

Moser, P. K. (2002). Cognitive idolatry and divine hiding. In D. Howard-Snyder & P. K. Moser (Eds.), *Divine Hiddenness* (pp. 120–148). Cambridge, MA: Cambridge University Press.

Muir, J., & Highland, C. (2001). *Meditations of John Muir: Nature's temple*. Berkeley: Wilderness Press.

Mulyana, A., Moeliono, M., Minnigh, P., Indriatmoko, Y., & Limberg, G. (2010, April). Establishing special use zones in national parks. *Center for International Forestry Research*. Retrieved from https://www.cifor.org/publications/pdf_files/infobrief/001-Brief.pdf

Nevin, J. A., Mandell, C., & Atak, J. R. (1983). The analysis of behavioral momentum. *Journal of the Experimental Analysis of Behavior*, *39*(1), 49–59. doi:10.1901/jeab.1983.39-49 PMID:16812312

Pais, A. (1979). Einstein and the quantum theory. *Reviews of Modern Physics*, *51*(4), 863–914. doi:10.1103/RevModPhys.51.863

Paleos.com. (2019). *Aristotle's Scala Naturae*. Retrieved from http://palaeos.com/systematics/greatchainofbeing/scala_naturae.html

Panksepp, J. (2007). Neuroevolutionary sources of laughter and social joy: Modeling primal human laughter in laboratory rats. *Behavioural Brain Research*, *182*(2), 231–244. doi:10.1016/j.bbr.2007.02.015 PMID:17363075

Pedersen, A., & Thomas, E. F. (2013). "There but for the grace of God go we": Prejudice toward asylum seekers. *Peace and Conflict*, *19*(3), 253–265. Retrieved from http://psycnet.apa.org/journals/. doi:10.1037/a0033738

Peterson, C. (2008, June 17). Other people matter: Two examples. Other people matter, and we are all the other people. *Psychology Today*. Retrieved from https://www.psychologytoday.com/us/blog/the-good-life/200806/other-people-matter-two-examples

Pettigrew, T. F. (1979). The ultimate attribution error: Extending Allport's cognitive analysis of prejudice. *Personality and Social Psychology Bulletin*, *5*(4), 461–476. doi:10.1177/014616727900500407

Piazza, J., Bering, J. M., & Ingram, G. (2011). Princess Alice is watching you: Children's belief in an invisible person inhibits cheating. *Journal of Experimental Child Psychology*, *109*(3), 311–320. doi:10.1016/j.jecp.2011.02.003 PMID:21377689

Plous, S. (2002). *Understanding prejudice and discrimination*. Boston, MA: McGraw-Hill.

Ramachandran, V. S. (2012). *The tell-tale brain: A neuroscientist's quest for what makes us human*. New York, NY: W. W. Norton and Company. doi:10.31338/uw.9788323529507

Ramachandran, V. S., Rogers-Ramachandran, D., & Cobb, S. (1995). Touching the phantom limb. *Nature*, *377*(6549), 489–490. doi:10.1038/377489a0 PMID:7566144

Sheldrake, R. (2005). The sense of being stared at--part 1: Is it real or illusory? *Journal of Consciousness Studies*, *12*(6), 10–31. Retrieved from http://citeseerx.ist.psu.edu/

Sternberg, R. J., & Sternberg, K. (2016). *Cognitive psychology* (7th ed.). Boston, MA: Cengage Learning.

Thomas, N. J. T. (2019). Mental imagery. In E. N. Zalta (Ed.), *The Stanford Encyclopedia of Philosophy*. Retrieved from https://plato.stanford.edu/archives/sum2019/entries/mental-imagery/

Wilson, E. O. (2016). *Half-earth: Our planet's fight for life*. New York, NY: Liveright Publishing.

Witte, R. S., & Witte, J. S. (2004). *Statistics* (7th ed.). Hoboken, NJ: Wiley and Sons.

ENDNOTES

1 In another chapter, the authors will compare this to the human belief in *representation* which does much the same; it isn't possible for many things to be represented by just one thing, but it isn't feasible to ask all of them either. The represented group is thus put into a category representing their average individuality, so that few are individually represented even if all seem to be represented as a heap. This is also true of the idea of "frequency" versus the diverse individual units of that frequency, in the field of neuronal dynamics (Gerstner et al., 2014).

2 Confirmation bias: a tendency to see new evidence as confirming of one's prior beliefs.

3 External goods consist of wealth, friends, and political power.

4 The Present is P, according to Hawking, the one event in spacetime that can be reached by whatever is part of that event (Hawking, 1998). So even if events in the past or future of P can affect P or be affected by P, nothing which is not part of the event can travel fast enough forward or back in space or time to participate beneath the event's horizon, nor can anything which is beneath that horizon ever leave it.

Chapter 7
Colonialism Disguised as Protection

ABSTRACT

Chapter 7 investigates the popular concept of environmental protection. The authors of this book timidly challenge the notion that nature needs human protection, since nature mainly needs protection from the human species. It seems more apparent that existing power structures use any vulnerable group to argue for the need to protect them, when what those disregarded groups want most is self-worth and unconditional affection, not to be kept guarded under lock and key. Often, the vulnerable species or immature stage of development said to be in dire jeopardy is not a stage at all, but tragically viewed as a stage along a ladder of conceptual merit from dependent to sovereign, and inedible seed to valuable fruit. This seems in evidence by a colonial worldview that sees juvenility as a waste of resources, rather than that brief liberty called childhood inherent to every species.

INTRODUCTION

"Every man I meet wants to protect me. I can't figure out what from" – Mae West

In this chapter, the authors will explain that the new popular demand to protect the environment is not motivated by a new, enlightened love of that environment. It would be terrific if it were, but there is too much evidence

DOI: 10.4018/978-1-7998-4408-2.ch007

that it isn't. Rather, the current call to protect the environment is still about exploiting it. The current situation is mostly about demanding dominion over living beings that have no means to represent themselves (which is how men of sufficient means and ambition have exploited women, children, and animals in the past; Goodell et al., 1985). But whether dominion is in the form of outright conquest and exploitation, or a subtle hegemony using the idea of extinction as a means to practice hegemony, it is still about who gets to control those who have no visible means of control. These authors ask the reader, why do living things need to be legally protected and intelligently controlled? How did they survive before humans came on the scene? If this service were offered *gratis*, at the unrecoverable expense of those devoted to saving Nature, one might be inclined to believe the present concern was a genuine act of altruism. But mostly one sees appeals for more money to fight against others making money off of Nature—even if Nature does not need money anywhere near as much as it needs mercy. It was the promise of money that drove Mankind to treat Nature as an unguarded pantry, and now it is money that promises to preserve what is left.

The authors submit that the idea of protection is not really about concern for another group. It is much more about a means to seize power, maintain power, and to justify the exacting of regular tribute to fund its exceptional claims of intragroup and intergroup dominance (Paulhus & Williams, 2002)[1]. The protector views itself as superior to the thing it protects, rather than humbling itself to what it says it cares about (Domination, 2019). This lack of simple reverence for what one boldly offers to protect has been called *paternalism* across history. Paternalism is defined as interfering with another's autonomy to benefit oneself—ostensibly to benefit the other at a cost of his or her autonomy (Goodell et al., 1985). One can also see a simulation of this demeaning relationship between a mind-like guardian and a body-like beast within each human individual, because the same mental phenomenon does the same thing to even one human being—wherein any ideas held dear by an individual may seek to influence, dominate, and exploit that individual in exchange for unrealistic promises of personal reward. This often surfaces as desperate obsessions and futile compulsions that an individual or a group of individuals are terrified of ignoring—lest the "chosen" entity lose its imagined favor, and incur the swift wrath of a jealous and outrageous misfortune (National Institute of Mental Health, 2013).

The authors will demonstrate that recent calls for protection of the environment are using the environment as a vulnerable group to further special interests. Across history those have usually been the *pecuniary interests* of one industry or another (adapted from *Barnes' Notes* on Acts 19:28, from Barnes et al., 1996). The authors will explain that biophilia should mean "I will love you no matter what," and not "I will love you so long as you do exactly as I tell you." The authors will talk about what beneficence is and what beneficence isn't. The authors will cite some examples of how the environment is being used to protect human jobs instead of Nature. The authors will talk about common misconceptions about fruition, maturation, and the strange view that the planet humans live on and its living, breathing inhabitants, are just lifeless resources that do not care how humans use them. The authors will argue that colonialism persists in the apparent defense of Nature, and a prejudicial, colonial worldview remains the predominant worldview, in that those who seek acceptance and protection must first acknowledge their inferiority to those that promise to protect them (Barnesmoore, 2018; Meinecke, 2018a, 2018b).

WHAT DOES IT MEAN TO PROTECT THE ENVIRONMENT?

Recent calls for protection of the environment are using the environment as a vulnerable group to solicit donations for the benefit of human industry. On any day, one can read appeals based on this disaster or that disaster, each of which are made by environmental rights groups (or human rights groups) who try to convince the public that more economics and more politics will help save the planet. Does this make sense? Wasn't it the quest for economic and political expansion that brought the planet to this juncture? There is an uncanny feeling of *environmental racketeering* in the practice of paying money for the protection of things that did not need protection before. It also tends to turn neighbor against neighbor, because some worry about losing their jobs if they sacrifice them to save the planet, while others wonder what good their jobs will do them if they let their planet die to save their jobs.

The authors submit that what is needed is not more politics; the authors submit that what is required is to let *natural selection* govern Nature once again—and let Nature do what it has always done best (govern itself). Politics is not a thing one sees in Nature; not many species have drones called lobbyists that try to influence drones called representatives on behalf of industrialists or special interest groups. Not many social species stop everything necessary to maintain life to hold debates wherein ambitious candidates ridicule one another to see which insect's platform is good for the economy.

The authors submit that offers of protection are not about love. Offers of protection are about careers—wherein some careers "protect people" and other careers "need protecting" (human jobs are about group economics, not group survival; Suzuki, 2007). Offers of protection seek acknowledgement that there are superior individuals who can protect inferior individuals said to be at risk otherwise. These offers are often associated deeply with narcissism, ethnocentrism, nationalism, Machiavellianism, and similar condescending attitudes toward other groups of beings (Sabet-Esfahani, 2014). The offer to protect a thing is motivated by the protector's belief that everybody needs it, not because the protector loves everybody. It is the requirement to acknowledge your weakness in exchange for its strength—not to acknowledge its unconditional compassion and adoration for you (Biernat, & Danaher, 2013; Glick & Fiske, 2012). Nature is fair game when it comes to gaming, because Nature does not know how to bluff, so it cannot tell when Men are bluffing[2].

This irony is captured in the social attribution known as *endearing yet inferior*, a paternal social attribution which has applied variously to the female gender, human offspring, the physically or mentally disabled, the chronically ill, the young, the elderly, the genetically inferior, and now the fragile environment as well (Glick & Fiske, 2012; Meinecke, 2017). Though humans know deep inside that what humans love always feels more amazing than they do (e.g. their wives, their children, and their planet), humans still think outwardly that they are more precious than what they protect (because humans can protect them from predators like themselves). Think about it: if humans think what they love needs their help to find the same honor in others it finds in them, are they complimenting it—or complimenting themselves? Do the concepts of awe and Majesty require superior intellect to discern? Yet, *intellect* is funny that way; it leads one to think it's basically humble, but it's anything but humble when bragging about its humility (Luke 18:9-12, KJV). The result of a possessive affection (rather than a pure affection), mental affection is frequently a form of jealous protectiveness and fear of loss, in place of a humble, grateful, reverent regard (Meinecke, 2018a, 2018b). Superior intelligence is often accompanied by a group of condescending, possessive urges—what psychiatrists call *morbid jealousy*—like the sudden need to monitor and control everything its beloved object happens to be doing and with whom it interacts (Somasundaram, 2010). Compare this to simply trusting what you love to engender tenderness no matter who encounters and falls in love with your beloved object. But if humans had just a little faith in what they loved, they would not need to watch over something they had faith in. *If she is lovely, she is always lovely*. She does not need you to be lovely. So long as somebody loves her, one can have faith she is being loved. Now apply this idea to the planet instead of a bride. The Earth, then, does not need humans to be lovely. The Earth just needs humans to show their faith in Her beauty again.

LOVE, NO MATTER WHAT

Protection is not about *love, no matter what*. Protection is about who comes first when push comes to shove (Sabet-Esfahani, 2014). For example, if it came down to the survival of the human species or the survival of the planet, which would humans prioritize? If it came down to saving a particular way of life or saving any way of life at all, which would humans prioritize? (The answer is easy; most would save their way of life and let the others vanish from the Earth). Better said, which needs protection more—the thing that protects, or *the thing* it protects? Obviously, what humans protect is what matters—not whom or what protects the beloved object. It is so odd that, if humans hold that a thing needs their special protection, humans tend to prioritize themselves (the protector) over what they protect. The most prominent example is the planet. What matters more—the survival of the planet or the survival of its protector? The planet can survive without the human species; it is humans who cannot survive without the planet and its species.

The problem here is obvious—if it came down to which to save, humans would save the protector and let the thing it protects perish (save Mankind instead of the Earth). The idea of protection is more about provoking worries where worries had not been before, which one sees broadly in the novel concern known as *identity theft* (Martin, 2019). A few years ago, this would have raised some serious concern. Suddenly a new terror of losing one's identity has created an emerging market offering to protect everyone's "identity." But that is only some general information about a living thing. No one can steal one's individuality. It is who you are. Since its inception just a short while ago, this idea of losing one's essence has now become a great economic niche (a positive thing for non-living industries) and a widespread personal anxiety (a negative thing for living individuals).

The idea of paying for protection is captured by definitions for protection, such as "the system of helping an industry in your own country" (by taxing the same industry of another country), or "the system of paying criminals so that they will not attack your business or property" (Protection, 2019). These scare the public into signing up to be protected (a very 20s and 30s era thing, when the economy was unstable, and money and jobs were scarce). However, humans do not need to scare one another into protecting the environment—from whom but the human species does the world need protection? Humans need only love their planet, and then it won't need anybody's protection. The argument that living breathing things need something else's protection is part of a dark triad of traits in which the predator projects aspects of himself as a means to frighten others into accepting his protection from those aspects (Furnham, Richards, & Paulhus, 2013).

The authors submit once again that protection is not about love, even if it claims to love what it fervently offers to protect. Protection is typically about nurturing a system of unfair exchange (because someone has to cover the costs of *running* whatever process will guarantee one's new need for safety). This additional administrative cost cannot be amortized away by its function alone, thus creating an escalating demand to help cover the cost of its administration. It is of special note that the very concept of a World Order is actually an unrecoverable expense… this is probably why it becomes more sacred over time than sustainable (Sabet-Esfahani, 2014).

Here are some examples of this offer to fill a non-essential need:

- In exchange for their protection, they promise to love you—meaning if you don't let them protect you, they won't love you, and will probably hate you as well for refusing them. A good example from English literature is Harold Skimpole's role in "Bleak House." His proposal to Esther Summerson is touching yet disquieting; when she repeatedly rejects his offers to lift her out of her station in life (not by much in any case) therefore under his protection, he becomes unexpectedly angry and jealous. Why would someone get mad at what he loves for not accepting an offer like that? Why wouldn't he be happy for her no matter whom she married—and rescued his beloved from their common misfortune? (Haynes, Mackie, Chadwick, & White, 2006).
- In exchange for your children, they promise to protect you (offspring are a common deal point when forming contracts with protective agencies, from Rumpelstiltskin to Abrahamic religions). Literature has many examples wherein children are bartered for social necessities. Perhaps one of the best and most recent is *Thousand Pieces of Gold*, in which a father is forced to sell his daughter so that the rest of his family

won't die of starvation (Kantor, Law, Sham, & Kelly, 1991). Money is the source of all sorrow.

Protection is a trope; it is part of an historically common ideology that views affection as a legitimate bargaining piece (protection in exchange for loyalty, fidelity, financial security, or affection; Coontz, 2004; Glick & Fiske, 2012). Down through human history, there runs a common thread in which a protective ideal eventually takes precedence over the living things it protects. This is especially common whenever humans defend ideological things that seem to need their help—even if internal concepts are not alive at all, just in danger of being forgotten (Damasio, 2003; Meinecke, 2017). Whenever a thing is weak or vulnerable, rather than making it stronger, humans argue that it needs their protection (so humans take away its freedom to keep it safe, in exchange for whatever it might contribute unwillingly to the group that protects it). By example, humans systematically domesticated vulnerable animals (that did not ask humans to tame them, and could not stop humans from taming them), in exchange for treating them as unwilling sacrifices to tutelary deities, as slave labor, and as food (Gilhus, 2006). This the intelligent creature imagines has improved the natural condition of unintelligent creatures by protecting them from much less voracious predators, or from starving to death (after humans seized all available food sources in their vicinity making that inevitable). Primitive tribes still exist that bring evidence of only a few human and animal victims a year (no social exchanges), compared to civilized societies in which untold numbers of both perish (based on social exchanges)[3].

If you have to pay someone to protect you, in criminal justice realms, this is called *racketeering* (not compassion or concern for their welfare). Often, sadly, payment comes in the form of adoption of beliefs that are unfair to anybody outside one's protected group, making you feel guilty for being in the protected group. Humans are a protected group, and humans are anything but fair to non-human animals (for example). To be a member of the human group, you must view animals as inferior (Costello, 2013; Costello & Hodson, 2014). But it does not stop there; not only do humans view groups outside their group as inferior, humans subdivide the human group into superior and inferior humans as well (Hodson, Kteily, & Hoffarth, 2014). Generally, to belong to the ingroup, you must accept the fact that all other groups are inferior.

GENUINE BENEFICENCE

Is protection motivated by the urge to be beneficent? A famous example is the concept of *beneficence* in the Belmont Principles (Borenstein, 2017). This is less about kindness as it is about maximizing benefit to the research group while minimizing harm to the subjects, so it is a curious word choice for this use (substituting "beneficence" for efficiency). Protection only seems beneficent because the alternative to being protected is worse; but genuine beneficence does not concern benefit, nor ever consider harm of any kind. The idea of a complete good (*pro bono*) does not propose inordinate options that carry consequences, some of which are less beneficent or might treat you as less equal when all things ought to be equal (since that implies coercion—no one chooses a worse option over a better one). Orwell (1964) had something to say on the subject of gradually inequal equality (that although all things are created equal, some things are more equal than others). Historically, the principle of beneficence implies a *universal good*, unusual charity (*sans* criteria), and a tender, *unexpected* mercy—not weighing the value of doing a bad thing anyway when humans are unsure if what humans are doing is "unreasonably evil" or "just a little evil" (Munyaradzi, 2012; Spurgeon, 2013).

How can one tell whether the motivation to preserve the environment is beneficent? Well, the main way to tell whether the motivation to protect something is for its sake (allocentric) and not for one's own sake (egocentric), is to see whether the one offering to preserve it expects anything in return. When conditions are present, conditions are the reason they are being done (otherwise, why have any conditions?). This is the basis of exchange theory, if one reduces it to a simple definition (Cook, Cheshire, Rice, & Nakagawa, 2013). If Nature must do something for humans before humans will do something for it, the fulfillment of that bargain takes precedence over doing what is good *for no reason at all* (for the good is usually no apparent reason at all). It's a standard part of forming social contracts. Otherwise, why would humans need contracts, if they could freely give one another what their beloved neighbor needed, with no contingencies to reach that gift of love?

For example, do humans refer to their environment as a "natural resource" (a means of supplying their urban environment with free goods), or do humans refer to it as Nature? The authors propose that Nature is more amazing than a gift of divine Providence, which purpose is to build human cities from its severed remains. It already has a purpose in its natural state, just as people do. If humans demand to own it, control it, watch over it, demand fealty from it, or demand tribute from it of any kind, their real reason for protecting it is not for some altruistic cause, but to demand its subordination in exchange for their superior guidance. At best, humans offer to protect something in exchange for paternal control, maternal control, or some hierarchical structure that will safeguard it if and only if it promises some sort of fealty to humans. But this is not love. This is not love of Life no matter what, and for no particular reason at all—other than the fact that Life of every kind is amazing! (Meyer & Bergel, 2012).

The view that wilderness and swamps are wasted land is a biased, uneducated view of amazing ecosystems teeming with life. History records a continuous view (by the human species) that anything it can conquer and control is fair game. Another name for this is *domestication*. Nothing domesticated is a companion. These are living things that work for humans so that humans do not see them as enemies; over time, humans forget that they forced them or tricked them into service. They begin to think that their relationship has always been warm and endearing (Coontz, 2004; Gilhus, 2006). But what reward do service animals receive as autonomous species in their own right? Consider the principle of justice (Borenstein, 2017). Do mice benefit from being used to discover cures for humans? Do turkeys (as a species) benefit from Thanksgiving? Many in the animal husbandry business think their food and clothing animals are actually better off as food and clothing, and argue that their wool would get too thick if humans stopped shearing them, or they would go extinct because nobody wanted to eat them. These authors wonder whether going extinct might sometimes be a species' means of emancipation from human exploitation, just as cell senescence may be a last-ditch cellular Masada against cancer (Zeng, Shen, & Liu, 2018).

The domestication or modification of something so that it cannot leave civilization (and survive) is neither beneficent nor loving. Love is not like that. Conditions on the granting of affection—implying that if humans hope to be fed and clothed and loved humans must be willing to commit acts against their own conscience—do not represent mutual affection. This is domination. This is social exchange theory. This is parental investment theory. This is a *conditional* bond that views what it loves as its property, and itself as its property's protector (Paulhus & Williams, 2002). But love isn't like that. Love does not see things as its property, but as its companion. Love does not see itself as a provider, but as a minister fortunate to be able to minister to the things it loves. When love is your motive, you don't domesticate whatever or whomever to a point where they are unable to survive without you. The authors submit that the idea of Providence is not a labor contract and the fear of breaching that contract, but a romantic bond of unimaginable tenderness and consolation (Beer, 1998).

The authors would like to clarify this with a bit of redundancy. The biophilic perspective is not about protecting things or nurturing things until they are old enough or wise enough to act like those who nurtured them; it is about the love of a living thing *no matter what*. The age or stage of life does not matter. Whether there is just one or thousands does not matter. Whether it seems to have developed perfectly or not does not matter. Whether it will survive for one day or for a hundred years does not matter. Is it alive? Then it is precious. Each day of life is precious, and no day is here so that "other days" can benefit from it. This is biophilia. It is a love of Life that does not see living things as potential rejects like a quality control inspector inspecting biological products. A grateful mother does not hold her "imperfect" baby in her arms for the first time and sigh, "If only my baby were perfect like *that* woman's baby over there" (Meinecke, 2017). What is more perfect than holding a new life in your arms that issued from your own womb?

What most think are imperfections, are often a justifiable opportunity for *more* time together, and a heightened attentional synchrony between mother and child (as many mothers of mentally challenged children have expressed). They are each one, *variations* on an amazing theme one can never see… never failures to meet some rare ideal of their kind (Darwin, 1876; Unterseher, Westphal, Amelang, & Jansen, 2012). A reverence for Life does not perceive living things as though they belong to something greater than Life, or that inferior living things belong to them (Meyer & Bergel, 2012). Wonder for Life does not rarely occur at some long-awaited (yet eerily temporary) state of fruition or decay. Wonder for Life happens *all the time* because one adores Life—all the time—no matter the stage, no matter the condition, so long as it lives and breathes. The reader may ask, does the need to breathe exclude plant life? Not at all—plants breathe too—but use something called *stomata* (tiny openings in leaves that open and close) instead of lungs. Respiration (breathing) is called *gas exchange* in both plants and animals, and each kingdom helps the other survive (Royal Society of Chemistry, n.d.).

So, when one sees living things as though they are in the process of "becoming" something, even though this sounds heart-warming, it isn't. What about what it *is* right now? How long must a creature wait to "be" something? Isn't it something all the time? Why must each living individual be the categorical fruit of something less categorical, or the indistinguishable product of something greater than itself? Has the reader noticed that humans actually spend more time worrying about the *survival* of a thing than simply adoring that thing right now? Humans view themselves as their *de facto* guardians, custodians, farmers, ranchers, and protectors. That sounds kind, but this is not love. This is an owner's view of his property, a caretakers' view of his charges, mistaken by one (or both) for an absent affection that would never think of seeing the other party as an object (Papadaki, 2010). Does it need their protection? Or does it need their undying affection? Should humans threaten one another with portents to elicit a universal mercy?

Perhaps, then, what is really needed is to stop "protecting the environment" and simply cherish it instead.

PROTECT THE FISHERIES BUT NOT THE FISH

Here is a real-world example in which popular demand to save the environment seems misguided. A recent call for action at Change.org highlighted the danger to an Alaskan Salmon fishery by allowing a mining interest to begin operations there (Dao, 2017). Curiously, the arguments against the mining operation were in defense of the current fishing industry (not what would happen to the fish or the bay itself). It labeled the fishing industry as a national treasure (rather than the bay itself with or without being used to exploit the fish there). It summed the crisis up, though, as the need to protect the salmon and beauty of Bristol Bay—when what it was really saying was "protect our jobs" in Alaska. It shared that Bristol Bay was the most valuable fishery and how many jobs would be lost if the bay were contaminated by runoff from the mining operations. It even went so far as to use the well-being of the indigenous people there as a reason to protect one industry from being spoiled by another industry (even if the *artisan* fishing traditions of the indigenous people had been displaced by the modern fishing industry, which is endangering the survival of fish worldwide). The call for action was not protecting the salmon from disaster (it is already too late for that). It was not preserving the natural beauty of Alaska (it is too late for that as well—wherever industry thrives in Alaska, Alaska's beauty has been lost). It was using these emotionally evocative cues to elicit contributions toward favoring the current fishing industry, not the wildlife—and even more curiously, new legal industries are cropping up because Nature needs a good lawyer as well (Earthjustice, 2019). The world humans live on does not need a lawyer. Who would dare to prosecute it—the People? It would indeed be curious to see a Supreme Court case of *The People vs. the Planet Earth.* Wild salmon do not live in a fishery. The call for protection, then, is not needed by the wildlife but by a human industry (commercial fishing). Before Bristol Bay was discovered by humans, the sockeye salmon of that bay had the best protection of all (humans had not yet discovered them).

As it turns out, environmental damage has been around for a very long time. As far back as 1289 CE, King Philip IV lamented the loss of watersheds and wildlife in his kingdom due to overfishing (Boissoneault, 2019). What did he do? He issued a proclamation to protect the kingdom's fishing industry (not the fish). Scientists actually note a "fish event horizon" around 1000 CE, when the needs of a burgeoning civilization transformed marine creatures and their families into a commodity rather than a delicacy.

JUVENILES, FRUITION, AND THE MYTH OF ENVIRONMENTAL PROTECTION

The authors mentioned this briefly in this chapter; this section will explore it a little more. The belief that something needs protecting often hinges on permission to exploit its weaknesses. Which governing entity exploits a weakness does not modify the idea of exploitation. However, the idea of "which governing entity" often constitutes a convenient way to argue for exploitation anyway—since humans hold that some governing entities are *less exploitative* than others and this is sufficient to justify their habit of exploitation. Even if some things seem more equal than others, *which* thing does a bad thing does not change the fact *that* a bad thing was done (Orwell, 1964).

The environment and its factors (soil, sea, sky, climate, plants, animals, birds) are very vulnerable to human predation, alteration (pollution), and exploitation. In fact, the Spanish word for enjoying the fruit humans gather from Nature (without fair compensation to Nature by the way), is *disfrutar*. The word means to garner joy by exploiting a thing. The word means treating it as "fruit" ripe for consumption instead of the offspring of its species, which it often is—although sometimes the flesh is sacrificial to induce involuntary transport of the seed. Still, humans do not defecate the way most species do, and so the consumption of fruit to transport the seed is denied that species by odd human attitudes toward natural body functions. The human idea of fruit is a curious fabrication that favors the eating of offspring over the rearing of offspring. As this author translated various Spanish language documents, it became clear that not all languages have words that cloak human exploitation of Nature using words that evoke warm, fuzzy feelings. Progressive nations and their languages have an inordinate number of abstract words that do not refer to Nature, and the words that survive are so denatured it is hard to argue against them to get back what they have replaced. Older languages sometimes carefully preserve (if not advertently preserve) their awareness of their attitudinal malpractices. In protecting the wording of their ancestors (which humans revere), humans often protect the history of their many abuses of wildlife as well (Meinecke, 2017).

Many words evolve until they are less descriptive of the living, growing things they referred to originally, as they are descriptive of their fitness for use as non-living municipal commodities. That fitness is aimed at growing the wealth of city-states, nation-states, and empire-states instead of simply growing (which is what living things do). Words like *juvenile* instead of child permit the civilized to think of a child without thinking about what childhood usually means—and what is being lost almost daily as humans commodify children at earlier and earlier ages (Deleuze, Guattari, & Foucault, 2009; Gray, 2013). Words like *fruition* allow the enlightened to think about different stages of the lifespan without thinking about how special each stage would be in its own right. They allow unchallenged participation in compulsory labor under a belief there is one ideal span of the lifetime when humans are fully human, when it is more probable it is a time when they are most valuable to industry as production workers (Fiske, 2011; Meinecke, 2017). Marchant (1646) once wrote, "What flees?" Then he went on to list the ages of a human lifespan (infancy, childhood, and so on). Oh how humans miss those things that flee. The authors ask the reader, then, why should a child feel guilty just because he or she is still a child? Is it a crime to be a child and only lawful to be an adult? Similarly, why should an injured or disabled or sick person feel guilty for being injured or disabled or ill? In both cases, they feel guilty because their productivity is impaired; even though the wish of every human is to be needed, these individuals feel that they are in everyone's way. They accuse themselves of being "less productive" than their conspecifics, and that because (in the abstract opinion of human systems), living things are here to preserve human concepts, not the other way around. This reversal of common sense judgment is not a new idea; it was mentioned two thousand years ago when Jesus wanted to heal someone on an "unlawful" day: "The Sabbath was made for man, not man for the Sabbath" (Mark 2:27, Matt. 12:11). But perhaps infirmity is not here so that some people may feel less valuable than others; perhaps infirmity is just a random chance for the healthy to humbly practice mercy, so that both healthy and infirm may feel equally valuable?

One sees children all the time in therapy who have not yet begun to live, yet pour their hearts out to a counselor saying how unfit they feel to exist. Society accuses them, and they accuse themselves, of not producing as much as they cost to maintain. What other species in Nature holds down a job, and which non-human parents wonder what their children will "be" when they grow up? This unnatural anxiety should not be added to a child's burden. "Am I loved?" should be the child's only concern. "Yes!" should not be far behind. But if humans spend their childhood worrying about some other stage of the lifespan, when will children revel in being children? This is how humans see trees (as lumber and wealth); this is how humans see animals (as food and wealth); this is how humans see their planet (as resources and wealth). But if you have one thing you *completely* love, why do you need more than one? A parent with one child prizes the child above every other (because the child has no siblings); a parent with many children experiences conflict in trying not to prize one child above another; a parent with no children is free of both anxieties—yet is frequently the most miserable, because even one child would deliver her or him from misery. It is the modification of anything one treasures that steals one's regard for it until it is modified (with reference to unconditional regard in Rogers, 1961).

THE FABLE ABOUT ENVIRONMENTAL GUARDIANS

When one takes a long, honest look, words like *juvenile* and fruition align with a specific kind of worldview—often called *Weltanschauung* in psychology (Colman, 2009, p. 817). For instance, many view a child as a kind of inferior human (not quite fully human, just ripening with a chance of becoming human). Not only that, they deem it a biological fact that everyone must view a child in this fashion, and the child that does not "develop" ought to feel ashamed of disappointing society after all society has invested in the child (psychologists call this juvenile onus *failure to thrive*). In some cultures, a child is not considered to be "human" or to have cultural value until the *bar mitzvah*, and up until that time, the son can be murdered without penalty because he is not legally human, and still a resource-consuming liability (as per the primary author's filicide research). Others conceive childhood as a social construction that is less biologically factual than sociologically factual. The need to process large numbers of unique humans is abetted by heaping humans by age. This helps to understand and measure their progress as to how to process them efficiently from just any old seed in the womb to a valuable drone in a cubicle—that perfect age between 18 and 65 when humans make good factory workers (Meinecke, 2017; Norozi & Moen, 2016).

The authors of this book propose that the view of a child (or any living thing for that matter), that has to go through a socially defined spectrum of discrete, developmental stages is an unjust bias of the dominant age group. A being that must be other things before it briefly is the one thing it was "always meant to be," represents an age-specific prejudice (ageism) against *all* of the epochs of that creature's lifespan that seem younger or older than that *desirable* stage of life. Curiously, this is also how humans judge between each member of a single species too (which member is most like that species, even if all of them must be *most like* that species—this is a requirement of any category). And if a living thing is only that desirable thing for a short while, while it is not quite that thing at most other times, how is "what it rarely is" a good definition of "what it usually is?" What does that say about Humankind's criteria for the value of Life at any stage when breath (and not criteria) determine what is or isn't alive?

This view is the view of Life as ripening fruit. The child is either not ripe enough, just right and ready to eat, or overly ripe and better off as animal feed. These are the same categories this author chose for his dissertation's method, because whenever human regard wanes sparse, these categories wax full (Meinecke, 2017). This *Weltanschauung* (opinion or attitude about the world) has viewed perfectly lovely women across the centuries as imperfect, whose only crime was infertility. Thus, a gorgeous living companion can be "put away" to make way for women who can bear fruit for the man who needs a rightful heir to his wealth. And perfectly good children can be made to feel depressed at the dawn of life, because they will not grow up to thrive and become materially successful (and thus meet their parents' expectations), especially after all that work in raising them (Fox & Bruce, 2004). Yet there are many examples of parents with "imperfect children," parents who invested the very same volume of time and resources in their children as the parents of "perfect children" (and probably a lot more). But instead of feeling saddled with an embarrassing product of their love for each other, the unusual bond between parent and child reminds one of the intrinsic value of Life itself. Human compassion often grows *where the red fern grows* when children are like strays (e.g. orphans or runaways or imperfect children); human banality often reigns otherwise, as perfect children easily succeed in overcoming their developmental hurdles (Rawls, 1961). It is of special note that animal scientists say the same thing about "mongrels" as opposed to pedigreed animals—mongrels make better companions and survive longer than those humans think are perfect, benefitting both animal and owner (Grandin & Johnson, 2005).

One might take a census of this perception of juvenility (a scorn of and avoidance of juvenility; Meinecke, 2017). Childhood is viewed as a stage of life when a human is not quite human yet. Does this make sense? But looking at the lifespan of plants the way humans do, has also resulted in *Homo sapiens* viewing its own children as developing plants. Society says children need society and need its protection, but with that argument society also denies them the free will to act independently (until they promise to deny others the same independence they are seeking). Civilized terminology is very telling here as well, as the authors noted earlier in this chapter about the arbitrary use of language to support one's current worldview. When a child thinks and behaves like those who held him or her captive for 18 years (that lifelong labor in exchange for belonging is normative and obligatory), society "emancipates" them. Isn't this akin to saying society will free a slave when that slave looks and thinks and acts like a slave owner? If a thing must stop being itself to be approved by society, how can one say society is open-minded and treasures individual differences? (Meinecke, 2017)

The authors note that the human species treats the natural environment the same way as it treats its children (except that natural kinds are unlikely to grow up and "become human," and thus will always need to be in protective custody). Yet the human species is pretty new on this old planet, and somehow all living species did very well without human "protection" before humans came on the scene. Perhaps then, the solution to environmental catastrophe (given that no other human solution seems promising), would be to restore the Earth to its former *absence of human authority*, and place human faith back in Creation (Nature). More protections will not help remedy what 5,000 years of protections have made infirm. The natural environment has already "grown up," and appears to be giving the human race signals that she (Nature) wants her independence from Civilization's so-called protection.

HOW COLONIALISM JUSTIFIES PREJUDICE

As it turns out, these fickle definitions that try to define childhood as an inferior stage of the human lifespan are not the only examples of fickle human nomenclature. Both religious experts and secular experts seem reluctant to refer to the plant and animal species (that humans exploit) as *children* even when referring to their obvious epoch of childhood. So, the experts substitute words like "juveniles" for both animals and human children, much as humans substitute words like "beef" for the corpses of cows. This practice of using a less obvious word to refer to things humans don't want to have pity for, conveniently dilutes the impact of what humans do to living things in their daily business affairs. In times past, humans referred to this differential view of the things they exploited (versus themselves for exploiting them) as *colonialism* (Barnesmoore, 2018; Meinecke, 2018a, 2018b). But humans don't think of the conquest and exploitation of the young as a form of expansionism (even if the evidence today is overwhelming, as childhood is treated as a nuisance species and education reaches earlier and earlier into infancy; Gray, 2013). This tendency to see juvenility as a liability is creating a myriad of issues for human children—who are struggling to figure out what must be wrong with them being born children instead of grown-ups. It is as though they should be ready to be employed at birth, so as not to be a burden on anybody who is already working (like their parents). The proliferation of childhood diseases corresponds very well to the proliferation of demands on their mental maturation long before the body is mature. G. Stanley Hall wrote of it one hundred years ago, when public education was seen far less reverently by a then more reverent and agrarian society (Search, 1901).

In fact, when it comes to *nameless children*, historians inform the reader that worshippers in ancient times were forbidden from using the actual names of their companion animals—when they were compelled to bring them to the temples and public squares to be sacrificed during the city's official bloodletting ceremonies (Gilhus, 2006). This nameless grouping of named individuals (as sacrificial lambs or offerings of human children) robs the members of that nameless group of their identity and individuality—and robs those who practice it as well of their conscience (Meinecke, 2017). But this helped the jubilant onlookers detach from pity for the hapless creatures who were being magically transformed into "meat" to feed the hungry masses (Gilhus, 2006). Similarly, the heaping of children into the category of "schoolchildren" helps adults watch their choiceless and depressing transformation from wayward little animals into controllable human sacrifices as well. Only at graduation do they get their names back, sort of like blue ribbons at the county fair, now hung on a cubicle wall from whence they eke out their best years making more sacrifices. It is odd that the animals humans bring to the fair and the children humans bring to their pageants and competitions both earn ribbons that neither of them really wear (or understand), but which are needed if they are to survive and reach safe pasture in their golden years.

In the same way, not only do humans *not* mention Bessie's name as they savor her remains on Sunday evening with some peas and mashed potatoes, they do not even call her a "cow" after she dies. (And humans call a human *a corpse* after they die too, robbing them at the very end of their one dignity). The consecration of a living companion for safe human consumption would be hard to blindly follow if everyone continued to think of her as a friend that they took care of just a few days before. These are things this author grew up witnessing with a blend of curiosity and horror, and which his wife carried inside as a lifelong trauma from a former marriage. The commercialization of living things requires the renaming of the things that humans consume, so they sound edible instead of friendly and helpless like a child. Bessie (who must not be named after she gets to market) is now just "a rump roast" or a stew with some nameless "beef" in it—not to exceed 15% fat which is bad for the consumer. And human children need honorary titles prefixed and postfixed to the names their parents so lovingly gave them at birth—before others will lovingly savor them as grown-ups. So curiously, even though it remains a good thing to receive those titles, they are not unique, but are shared by everyone else who has that title—whereas a given name is often unique to that individual, even if it implies no specific honor.

When it comes to children, the news does not often say that helpless children are being imprisoned against their will, or that juvenile delinquents or illegal immigrants and their offspring are being detained without bail (Stevenson & Stinneford, 2020). Society does not compare its own conduct to those society condemned in the past for such conduct—instead, society accuses those they detain in cages of making "poor choices." The environment is just the newest in a long line of guarantees whose protection humans need to guarantee. Yet if it needs to merit society's approbation to find the same justice as society, perhaps it is society's perception of approbation which is incomplete or unnecessary (Meinecke, 2017).

CONCLUSION

In this chapter, the authors explained protection of the environment is more about power than pity. Calls for protection of the environment are using the environment as a vulnerable group to further special interests such as the commercial fishing industry. The authors explained that biophilia (love of life) ought to mean *love no matter* what—when so often it seems even *love comes at a cost* few can afford these days. The authors talked of a principle that science calls beneficence but oddly defines as a maximizing of Mankind's scientific gains while minimizing Nature's biological losses—so that scientific exploitation of Nature, not Nature, remains sustainable. The authors spoke of common misconceptions about the fruition of living things which deny those things the unconditional regard they need from Humanity lifelong (Rogers, 1961). The authors argued that colonialism persists in the guise of concern for the environment and a fallacious need for its continued protection by the very groups that exploited it. And lastly, the authors compared the persisting colonial worldview to a mindset that expects that all those who seek its protection must acknowledge their inferiority first (Barnesmoore, 2018; Meinecke, 2018a, 2018b). At this juncture in human history, it is also of major concern whether this species perceives the colonization of space the same way it perceived the conquest of its planet, leaving nothing anywhere untrammeled by its insatiable quest for domination and modification (Deleuze et al., 2009).

REFERENCES

Barnes, A., Murphy, J. G., Cook, F. C., Pusey, E. B., Leupold, H. C., & Frew, R. (1996). *Barnes' notes on the Bible*. Grand Rapids, MI: Baker.

Barnesmoore, L. (2018). *Worldview, order and human-nature relations: Geddes, Mumford, modernity and a true return to nature*. Retrieved from https://www.academia.edu/

Beer, J. B. (1998). *Providence and love: Studies in Wordsworth, Channing, Myers, George Eliot, and Ruskin*. Oxford: Oxford University Press. doi:10.1093/acprof:oso/9780198184362.001.0001

Biernat, M., & Danaher, K. (2013). Prejudice. In I. B. Weiner (Ed.), Handbook of psychology (Vol. 5, pp. 341–367). Hoboken, NJ: Wiley.

Boissoneault, L. (2019, May 21). *The medieval practices that reshaped Europe's fish*. Retrieved from https://www.theatlantic.com/science/archive/2019/05/medieval-people-were-already-ruining-fish/589837/

Borenstein, J. (2017). Research involving human subjects. *CITI Program*. Retrieved from https://www.citiprogram.org/

Colman, A. M. (2009). *Oxford dictionary of psychology* (3rd ed.). New York, NY: Oxford University Press.

Cook, K. S., Cheshire, C., Rice, E. R., & Nakagawa, S. (2013). Social exchange theory. In *Handbook of Social Psychology* (pp. 61–88). Dordrecht: Springer. doi:10.1007/978-94-007-6772-0_3

Coontz, S. (2004). The world historical transformation of marriage. *Journal of Marriage and the Family*, *66*(4), 974–979. doi:10.1111/j.0022-2445.2004.00067.x

Costello, K. (2013). *Determinants and consequences of dehumanization: An interspecies model of prejudice* (Doctoral dissertation). Brock University, St. Catherines, Ontario, Canada.

Costello, K., & Hodson, G. (2014). Explaining dehumanization among children: The interspecies model of prejudice. *British Journal of Social Psychology*, *53*(1), 175–197. doi:10.1111/bjso.12016 PMID:23083220

Damasio, A. (2003). *Looking for Spinoza: Joy, sorrow, and the feeling brain*. Orlando, FL: Harcourt.

Dao, T. (2017). *Protect Bristol Bay from open pit mining*. Retrieved from https://www.change.org/p/epa-protect-bristol-bay-from-open-pit-mining

Darwin, C. R. (1876). *The origin of species by means of natural selection, or the preservation of favoured races in the struggle for life* (6th ed.). London, UK: John Murray.

Deleuze, G., Guattari, F., & Foucault, M. (2009). *Anti-Oedipus: Capitalism and schizophrenia*. New York: Penguin.

Domination. (2019). In *Oxford's Learner's Dictionaries online*. Retrieved July 30, 2019 from https://www.oxfordlearnersdictionaries.com/us/definition/english/domination

Earthjustice. (2019). *Alaska's Bristol Bay and the pebble mine*. Retrieved from https://earthjustice.org/features/alaska-s-bristol-bay-the-pebble-mine

Fiske, S. T. (2011). *Envy up, scorn down: How status divides us*. New York: Russell Sage Foundation.

Fox, G. L., & Bruce, C. (2004). Conditional fatherhood: Identity theory and parental investment theory as alternative sources of explanation of fathering. *Journal of Marriage and the Family*, *63*(2), 394–403. doi:10.1111/j.1741-3737.2001.00394.x

Furnham, A., Richards, S. C., & Paulhus, D. L. (2013). The Dark Triad of personality: A 10-year review. *Social and Personality Psychology Compass*, *7*(3), 199–216. doi:10.1111pc3.12018

Gilhus, I. S. (2006). *Animals, gods, and humans: Changing attitudes to animals in Greek, Roman, and early Christian thought*. London, UK: Routledge. doi:10.4324/9780203964798

Glick, P., & Fiske, S. T. (2012). An ambivalent alliance: Hostile and benevolent sexism as complementary justifications for gender inequality. In J. Dixon & M. Levine (Eds.), *Beyond Prejudice: Extending the Social Psychology of Conflict, Inequality and Social Change* (pp. 70–88). New York, NY: Cambridge University Press.

Goodell, G. E., Aronoff, M. J., Austin, D. J., Cadeliña, R. V., Emmerson, D. K., Hansen, K. T., ... Sinha, A. C. (1985). Paternalism, patronage, and potlatch: The dynamics of giving and being given to [and comments and reply]. *Current Anthropology*, *26*(2), 247–266. doi:10.1086/203252

Grandin, T., & Johnson, C. (2005). *Animals in translation: Using the mysteries of autism to decode animal behavior*. New York, NY: Houghton Mifflin Harcourt.

Gray, P. (2013, September 18). *The play deficit*. Retrieved from https://aeon.co/magazine/being-human/children-today-are-suffering-a-severe-deficit-of-play/

Haynes, S., & Mackie, L. (Producers), Chadwick, J., & White, S. (Directors). (2006). Bleak house [Mini-series]. United Kingdom: BBC.

Hodson, G., Kteily, N. S., & Hoffarth, M. R. (2014). Of filthy pigs and subhuman mongrels: Dehumanization, disgust, and intergroup prejudice. *TPM. Testing, Psychometrics, Methodology in Applied Psychology*, *21*(3), 267–284. doi:10.4473/TPM21.3.3

Kantor, S., Law, L., & Sham, J. (Producers), & Kelly, N. (Director). (1991). *Thousand pieces of gold* [Motion picture]. USA: Greycat Films.

Marchant, J. (1646). *Rationale evangelizantium, sive doctrina et veritas evangelica, a sacerdotibus, pastoribus, concionatoribus, pectori appendenda, plebique per anni circulum è Cathedris proponenda, exponenda: Duobus tomis comprehensa*. Paris, France: Apud Michaelem Soly.

Martin, G. (2019). *The meaning and origin of the expression: identity theft*. Retrieved from https://www.phrases.org.uk/meanings/identity-theft.html

Meinecke, L. D. (2017). *Neglected by assessment: Industry versus inferiority in the competition for scarce kidneys* (Doctoral dissertation). Available from ProQuest Dissertations and Theses database. (ProQuest No. 10689852)

Meinecke, L. D. (2018a). A theory of cognitive idolatry, or the human struggle to preserve psychological certainty in a biological species - chapter 1 (Research proposal). doi:10.13140/RG.2.2.16527.43684

Meinecke, L. D. (2018b). A theory of cognitive idolatry, or the human struggle to preserve psychological certainty in a biological species - chapter 2 (Research proposal). doi:10.13140/RG.2.2.31089.53601

Meyer, M., & Bergel, K. (2012). *Reverence for life: The ethics of Albert Schweitzer for the Twenty-First Century*. Syracuse, NY: Syracuse University Press.

Munyaradzi, M. (2012). Critical reflections on the principle of beneficence in biomedicine. *The Pan African Medical Journal, 11*(1), 29. doi:10.4314/pamj.v11i1 PMID:22514763

National Institute of Mental Health. (2013). *Obsessive-compulsive disorder*. Retrieved from https://www.nimh.nih.gov/health/topics/obsessive-compulsive-disorder-ocd/index.shtml

Norozi, S. A., & Moen, T. (2016). Childhood as a social construction. *Journal of Educational and Social Research, 6*(2), 75. doi:10.5901/jesr.2016.v6n2p75

Orwell, G. (1964). *Animal farm*. New York: Harcourt, Brace and Company.

Papadaki, L. (2010). What is objectification? *Journal of Moral Philosophy, 7*(1), 16–36. doi:10.1163/174046809X12544019606067

Paulhus, D. L., & Williams, K. M. (2002). The dark triad of personality: Narcissism, Machiavellianism, and psychopathy. *Journal of Research in Personality, 36*(6), 556–563. doi:10.1016/S0092-6566(02)00505-6

Protection. (2019). In *Oxford's Learner's Dictionaries online*. Retrieved July 22, 2019 from https://www.oxfordlearnersdictionaries.com/us/definition/english/protection

Rawls, W. (1961). *Where the red fern grows*. Bantam.

Rogers, C. R. (1961). *On becoming a person*. Boston, MA: Houghton Mifflin.

Royal Society of Chemistry. (n.d.). Gas exchange. In *Chemistry for Biologists*. Retrieved from https://www.rsc.org/Education/Teachers/Resources/cfb/gas.htm

Sabet-Esfahani, S. (2014). *Prejudice and protectionism: Essays at the intersection of international political economy and psychology* (Doctoral dissertation) Harvard University.

Search, P. W. (1901). An ideal school, or looking forward. In W. T. Harris (Ed.), *International Education Series* (Vol. 52, pp. 1–357). New York, NY: D. Appleton and Company.

Somasundaram, O. (2010). Facets of morbid jealousy: With an anecdote from a historical Tamil romance. *Indian Journal of Psychiatry*, *52*(3), 284–288. doi:10.4103/0019-5545.71007 PMID:21180423

Spurgeon, C. H. (2013). *The salt-cellars: Being a collection of proverbs, together with homely notes thereon*. Harrington, DE: Delmarva Publications.

Stevenson, B. A., & Stinneford, J. F. (2020). *The eighth amendment*. Retrieved on March 16, 2020 from https://constitutioncenter.org/interactive-constitution/interpretation/amendment-viii/clauses/103

Suzuki, D. (2007). *The sacred balance: Rediscovering our place in nature, updated and expanded*. Vancouver, BC: Greystone Books Ltd.

Unterseher, M., Westphal, B., Amelang, N., & Jansen, F. (2012). 3,000 species and no end–Species richness and community pattern of woodland macrofungi in Mecklenburg-Western Pomerania, Germany. *Mycological Progress*, *11*(2), 543–554. doi:10.100711557-011-0769-7

Zeng, S., Shen, W. H., & Liu, L. (2018). Senescence and cancer. *Cancer Translational Medicine*, *4*(3), 70–74. doi:10.4103/ctm.ctm_22_18 PMID:30766922

ENDNOTES

1 If one traces the beginnings of dominance claims, they often correspond to an etiology of thinking not unlike a *dark triad* of traits symbolic of a verbal disease. This disease spreads by infecting vulnerable hosts in desperate need of esteem with stories about their greatness (deception),

lack of remorse for their actions (psychopathy) and a belief that they alone hold the exclusive traits needed to help everyone survive (narcissism). It is argued here that language is much less an intellectual tool than a simple form of transport for chemically rewarded, verbally transmitted diseases (hosted subliminally as intrusive, intractable schema). Like any parasite that gains entry using pathways meant for biologically beneficial substances, "words" (sacred patterns of sound) infect their hosts with biologically malevolent urges, similar to airborne illnesses—except this pathway requires speech comprehension and production to reproduce and spread among a group of vulnerable and unwary hosts (Berns et al., 2012; Meinecke, 2018a, 2018b).

2 Bluffing is the act of deceiving someone into believing you can or will do a thing.

3 This was the WVC anthropology and religions class the other took circa 1975 – the professor had home movies of it, and we watched them. It made the author challenge his beliefs about our societies and our belief in our nobility.

Chapter 8
At the Crossroads:
Sovereignty or Faith?

ABSTRACT

The theme of Chapter 8 encompasses the current crisis and the theory that human dominion is killing the planet and humans with it rather than protecting it. Denial of climate change and the belief that nature needs human intervention to survive instead of the reverse suggest the species is not fit to be the steward of its world. Human beings seem to view the conquest and extirpation of nature as a sort of progressive and inevitable victory over their savage past . . . even though the human species was once in awe of the natural world. These days, the species hominin seems more savagely eager to transform that awe into salable goods and global power. Perhaps the answer to saving the planet is not more sovereignty, but more faith.

INTRODUCTION

Dominion and Decision-Making

Humans tend to believe that they are excellent decision-makers, and that they have the right to assume dominion over the Earth and all its inhabitants (animals, plants, land, water, natural resources, etc.). They seem to have developed a case of *biophilic amnesia* because, when left undisturbed, non-humans are completely capable of maintaining Nature's balance by utilizing a number of innate strategies that include shifting the gender ratio when either

DOI: 10.4018/978-1-7998-4408-2.ch008

males or females become too numerous within a particular species, or by resorting to asexual reproduction when necessary to guarantee a population's survival for generations to come. In most species, the sex ratio tends to be 100 males for every 100 females; however, in the human species, the sex ratio is biased toward the male gender, which is about 105 males for every 100 females. Males have a higher risk of dying due to the sex differentials in natural death rates (risk-taking, probably) as well as from external causes, such as wars, accidents, violence, and injuries. In cases where a country exhibits a preference for males, but its sex ratio does not equalize or exceeds the 105:100 ratio, society will intervene to reduce the population of females through infanticide and/or selective abortion (WHO, 2019).

Non-human species, such as animals and plants, display a unique capacity to maintain control of their numbers by permitting an ample supply of predators within a territory, so that their food supply remains within manageable limits. Problems tend to arise when human populations decide to encroach upon undeveloped areas teeming with wildlife. Humans tend to perceive wildlife as natural resources for their personal harvesting, a threat to their agricultural and animal husbandry enterprises, or a nuisance species that needs to be eradicated or relocated. In the absence of human interference, usually there is no need for wildlife "management" since ecosystems evolve some degree of self-regulation; however, since virtually all ecosystems have been significantly impacted by some type of human activities (the expansion of agricultural pursuits, logging, mining, and urbanization, all of which result from human population growth), wildlife management has become a necessary intervention. Game animals are the initial casualties of natural resource depletion as a means of feeding workers or to provide an alternative income stream for workers, or as collateral damage when other natural resources (timber, minerals, etc.) are being exploited. As a result, wildlife management becomes an absolute necessity in an effort to prevent the overexploitation and/or extirpation of the existing species needed to provide for a sustainable human economy (Stokes, 2011). What is of most note, is the motivation to curtail exploitation (primarily for human interests in their sustainability as exploitable species, not out of compassion or s sense of conscience).

Game Harvesting can be defined as the taking of wildlife by hunting or trapping. This is analogous to the process of farmers harvesting their crops or plants where game are harvested for several reasons:

1. The yield of valuable products, such as furs, hides, tusks, teeth, bones, meat, etc. for use in traditional ceremonies or medicines. In South Africa, endangered vultures are harvested illegally for use by local witch doctors or traditional healers for magical purposes, including successfully playing the lottery, since these vultures are believed to have the ability to see the future (and winning lottery numbers). This was practiced by the Romans to predict the outcomes of important decisions also, and was once called *augury* (Hunt, 2007).
2. As rites of passage and proof of "manhood", for example in western Kenya and Tanzania where killing lions was required for young Masai men to achieve warrior status in their tribe. Such biologically violent rites (to acquire psychosocial acceptance) were also popular among native American tribes (Smithee, 2019), and are still cultural rites of passage in developed nations (killing one's first deer).
3. The reduction of human-wildlife conflict by government-supported eradication programs, such as the establishment of bounties on coyotes by the United States government during the early twentieth century. Rather than eradication, this resulted in the evolution of more intelligent and adaptable coyotes which have now spread throughout the entire continental United States—when they had been only living in the central and western states originally. It is of note that similar efforts to eradicate criminal behavior have also resulted in more intelligent criminals.
4. The mitigation of bird strikes with aircraft by the enactment of shooting programs at many United States airports, such as JFK International in New York City. In excess of 63,000 laughing gulls (not at all humorous) were shot during the decade of the 1990s according to the U.S. Fish and Wildlife Service data (Dolbeer, Chipman, Gosser, & Barras, 2003).
5. Since the turn-of-the-century, *ecotourism* (particularly wildlife ecotourism) has resulted in a new valuation of non-consumptive game resources, such as wildlife photography, diving in shark cages, so-called "green" hunts or darting safaris in which game animals are tranquilized in contrast to being killed for trophies, and paid participation in wildlife research via organizations akin to *Earthwatch.*
6. In rural regions that offer poor economic opportunities such as technology or large cities, economic reasons drive the management of wildlife by governments and/or government agencies through the idea of "outdoor sportsmen". Using the sale of hunting, trapping, and fishing licenses as well as the taxing of hunting equipment and other expenditures specific to hunters, these regions are able to sustain their rural existence without

capitulating to urban development (by means of revenue from animal exploitation management). There is also a federal excise tax on hunting equipment in the United States that includes bows and ammunition. In 2009, excise taxes produced almost $473 million in apportionments that were returned to states in 2010.

7. Economic interests also drive the interest of the private sector in the management of wildlife resources. For example, in post-apartheid South Africa, ecotourism and hunting have resulted in the proliferation of private hunting ranches and game preserves. These have become a significant source of income, and now represent a greater land area than the public reserves. As a non-human benefit, though, much natural habitat is also preserved which protects several non-game species simultaneously (Stokes, 2011).

THE PHILOSOPHICAL MODELS OF WILDLIFE/ GAME MANAGEMENT PRACTICES

Although there are several historical philosophies underlying the philosophies of wildlife management, the often-cited *dominion model* is based upon the premise that humans have the right to exploit all of the planet's natural resources in any way they choose as well as in any quantity that they deem. This can be found in the traditional Abrahamic religions where humans are believed to not only *have* absolute dominion over Nature, but be *charged* with its subjugation (which seems illogical, since if one has been given dominion one should not also have to achieve it). The *royal* or *regal model* assumes a similar perspective in which a ruling individual or entity (a squire for example) has the divine right to claim the ownership of wildlife resources, and to declare the "poaching" by unauthorized persons of those resources to be illegal hunting (Gallagher, Hogg, Marmion, & Trickelbank, 2009). These models often were more focused on depriving other humans of favored resources rather than for wildlife management proper (making it illegal for the poor to eat meat, essentially). The third philosophy is the *colonial model*, which expanded the control of the regency to include colonial holdings—which often resulted in wildlife populations becoming private shooting ranges for colonial officers as well as a means of disenfranchising the indigenous tribes and depriving the colonized individuals of the best resources. In response to these models of mismanaged wildlife populations, a *managerial model* evolved in which certain species are specifically managed for regional socioeconomic or sport purposes, such as for anglers who prefer fishing for trout and salmon, as well as for the pen-raising of game birds for release on shooting preserves. Another reaction to mismanaged wildlife resources gave rise to the *legal* approach where strict laws were enacted to "protect" wildlife (from unauthorized exploitation only), in contrast to prior royal mandates which functioned similarly, enforceable by a public agency charged with the responsibility to selectively prohibit hunting (Gallagher et al., 2009). Traditional societies can also have informal laws or rules that are socially enforced to selectively manage local wildlife resources, often under the guise of concern for the animals and their ecosystems, but without actually removing their sovereignty over them (Stokes, 2011).

The twentieth century witnessed the rise of *ecology* as a science promoting a global awareness of environmental ethics together with the emergence of the multidisciplinary science of the field of *conservation biology*, which evolved into an *ecological philosophy* of wildlife management. This ecological perspective for wildlife management promotes the rights of other species to exist as well as the interconnectedness of species, and the recognition of their resources which therefore values all wildlife, not just the economically valued game species. This ecological perspective usually focuses on the management of habitats and ecosystems rather than on individual species, unless those species are categorized as being heavily harvested and/or endangered (Stokes, 2011) which clearly aligns itself with The Theory of Biophilia (O'Grady, 2016). Notwithstanding that alignment, the authors explained in the previous chapter how even this seemingly positive advent has morphed into another form of colonialism wherein the vulnerable species are used mainly to solicit empathy from an otherwise powerless population, and raise funds for primarily political ends.

HABITAT AND ECOSYSTEM MANAGEMENT

The burgeoning human population often finds itself in conflict with other burgeoning non-human populations when both types of populations attempt to violate the *Pauli Exclusion Principle* (Quinn, 2015) by trying to occupy the same space or territory at the same time. This conflict is further exacerbated when the human population has managed to eradicate competing apex predators who had been deemed "dangerous" to humans or their property—although they had been keeping their "nuisance" prey within reasonable limits.

Although this lower-on-the-food-chain prey typically present no direct danger to humans, they can present some issues when their unchecked populations decimate crops, or infiltrate human abodes thereby competing with humans for food and shelter. Human beings have frequently attempted to resolve such non-human populations issues through the introduction of non-native species into a particular geographical area, often with unanticipated and unintended results. Not only does introducing an alien species into an area that has no indigenous predators for population control not solve the "perceived" existing problem, but it also creates future problems when non-native species reproduce unchecked, and become a major cause of decreased biodiversity.

The disastrous outcome of human meddling is evidenced by Florida's experiences with the introduction of non-indigenous species of plants and animals, such as the Melaleuca trees that were introduced to Florida in 1996 from Australia to drain the Everglades (Christensen, 2018). Unfortunately, little or no concern was aroused for the future impact that this draining of the Everglades (aka the "river of grass"), for the convenience of building developers, would have on Florida's climate and wildlife. Florida's ecosystem is extremely water-oriented since the state is a peninsula having an extensive canal system for the purposes of drainage and transportation. So, again with a lack of aforethought, when Chinese grass carp were deliberately introduced into many of Florida's lakes and ponds in 1972, it seemed like an inexpensive, biological control method to address issues being caused by invasive aquatic plants that had taken root there. However, the policymakers soon learned that these non-native fish would grow up to 50 inches in length and weigh up to 75 pounds in their transplanted environment. As of 2010, biologists had also discovered that Chinese grass carp displayed an uncanny ability to outsmart every type of fishing technique that had been designed to successfully remove them from Florida's water systems (Chinese Grass Carp, 2019). An intriguing aspect of these intelligent interventions is the failure to acknowledge the cleverness of animals who do not wish to die or be used for human purposes. This aligns with the human species' belief in its exceptionalism.

To further clarify terminology, *alien species* can be described as species that are considered to be *exotic, introduced, non-indigenous*, and/or *invasive* which do not belong to the ecosystems in which they are intentionally or unintentionally introduced. Alien species can consist of, but are not limited to fish, plants bacteria, viruses, crustaceans, and/or mollusks, and tend to disrupt the balance of an ecosystem through their rapid reproduction. Trade, which is the hallmark of human encroachment and species-wide pride, is perhaps one of the widest breaches in previous population boundaries—that once limited the unchecked propagation of devastating diseases, plants, and animals to natural (non-intellectual) barriers. Among the vectors transporting alien species into marine ecosystems are the commercial shipping vessels discarding their ballast water, as well as humans discarding formerly captive animals and/or plants into freshwater estuaries, rivers, lakes, and streams. Well-known invasive species include the following: the Northern Snakehead (a fish), the Zebra mussel, the Sea Lamprey, and the Asiatic Clam. The introduction of alien species exerts a number of negative effects on human health and well-being, such as increased predation and competition for food, shelter, and habitat; introduction of diseases; extinction of native plants or animals; and genetic alterations in populations (MarineBio, 2019).

The introduction of alien species can be intentional or unintentional, but the negative impact(s) on the unsuspecting ecosystems can be equally disastrous. For example, unintentional introduction is exemplified by the invasive algae *Caulerpa* or the ornamental plant, the Purple Loosestrife or the Purple Star Thistle have been released by humans into local waterways where they quickly overtake and choke out native plants as well as disrupt the water flow in lakes, rivers, streams, and estuaries. Unwanted exotic fish, such as the venomous red lionfish, *Pterois volitans* have literally invaded the waters around the Southeastern United States (MarineBio, 2019). In contrast, intentional introductions of alien species included the toxic cane toad into Australia in 1935 in an effort to control the native grey-backed cane beetle and the French beetle, which has resulted in a rippling effect throughout Australia's ecosystem rarely experienced with the spread of an invasive species—a *trophic cascade* (Slezak, 2015). When the intentionally introduced animal multiplies very quickly and assumes control of its non-native habitat and beyond, the *best laid plans of mice and men* can backfire in ways never imagined (Burns, 1785). This is evidenced by the introduction of the mongoose into Hawaii, which was originally intended to control the rat population on the island; however, it eliminated many of the native bird species in the process. A government-sponsored program to control erosion in the southern part of the United States has resulted in the indefatigable kudzu plant blanketing much of the area. Even the development or construction of new seaways, canal systems, cross-basin connections, etc. linking bodies of water can provide a means for alien species to migrate into novel territories as exemplified by the invasion of the Great Lakes following the construction of the St. Lawrence Seaway in 1959 (MarineBio, 2019). In each case, one sees a novel involvement of ecologically neglectful "rational" processes, into what were previously self-limiting, purely biological systems—usurped by the belief that the human mind can do better in the short term than evolution could accomplish in millions of years (followed by unforeseen biological tragedies).

Vector ecology has developed from research conducted to determine how alien species are introduced into a native system, gain a better comprehension as to why certain species flourish in non-native environments, and to determine the impact(s) they are having on the native species. Studies in *biogeography* provide important data and information regarding alien species' global distribution patterns, while organized databases enable researchers to collect and analyze that data and information to develop future strategies to mitigate and/or eliminate the introduction of alien species (MarineBio, 2019). So, while humans *may* be learning from their erroneous decision-making and mistakes, how much better might it be if they had not made them in the first place? How much better might it be if humans just allowed Nature to evolve on her own terms instead of making decisions in areas where they have little or no knowledge regarding the long-term or even the short-term effects of "fooling" with Mother Nature? This ardent and misguided perseveration (the use of more rational agents to solve the disasters caused by the last team of rational agents) suggests a neglect of the primary stakeholders (whom are non-rational agents) in the non-anthropocentrically driven struggle for existence (Darwin, 1876).

HABITAT DESTRUCTION AND PLANETARY POLLUTION

A quote by Greg Griffin reiterates the Dominion Theory quite succinctly: "So much of the habitat destruction and pollution is based on the simple principle that we somehow have been given free license over other species to degrade the planet," (Madaan, 2019, p. 1). Humans are the most wasteful and destructive species on the planet—as evidenced by the number of creative ways in which they have polluted the Earth. In general, *pollution* is categorized according to the area of the environment which is negatively impacted, such as air, water, land, soil, thermal, light, industrial, and visual; typically those negative impacts overlap into multiple areas, for example land *and* soil. Typically the types of pollution with which most people are familiar are air, water, and land; however, the remaining five categories also make major contributions to the degradation and destruction of the Earth (Madaan, 2019).

It is of especial note that in the popular book and television series *Good Omens*, one of the four horsemen of the apocalypse (conventionally called

Pestilence in the book of Revelations) was renamed Pollution (Gaiman & Pratchett, 2019). The Greeks held that defamation of sacred things and unclean habits polluted otherwise holy things. Human expansionism is certainly representative of such an attribution, the defiling of the planet while viewing it as subhuman or demonic, and bringing into existence demons of unimaginable banality (Hodson, MacInnis, & Costello, 2014; Moses & Stone, 2013).

Air has become the most polluted environmental resource, and consists of the introduction of harmful substances into the air which diminishes the air quality to where it is detrimental to the health and well-being of humans as well as non-human species including plant life who depend upon it for their very survival. Air pollution occurs when foreign or toxic substances, such as fumes, odors, dust or particulates, and foreign gases are released into the air by either humans or natural activities at levels which adversely impact the comfort and/or health of humans and non-humans. These pollutants are generated by manufacturing industries, power plants, vehicles, smoking, wildfires, volcanic eruptions, and the burning of waste materials (plastics, rubber, etc.), and are comprised of such compounds as hydrocarbons, carbon monoxide, sulfur oxides, volatile organic compounds or VOCs, chlorofluorocarbons or CFCs, and nitrogen oxides. Dust and particulate matter are also classified as common air pollutants (Madaan, 2019).

Water pollution can be defined as the contamination of bodies of water, such as oceans, rivers, lakes, streams, aquifers, ponds, groundwater, etc. which occurs when toxic or harmful foreign substances, such as raw sewage are introduced either directly or indirectly into those bodies of water. *Any* alteration(s) in the chemical composition, biological or physical properties of water are categorized as water pollution. Often humans are the primary contributors in the pollution of water with toxins and harmful chemicals, and the origin of the human activities which pollute the water are classified as *point source* or a single/identifiable source, *non-point source* or the cumulative effects of multiple contaminants, and/or *groundwater.* The latter type of pollution occurs via infiltration of wells, aquifers and other groundwater sources. Following air pollution, water pollution is deemed to be the second most polluted environmental resource on the planet. It is disturbing to realize that the (preventable) sources of water pollution include:

- Discharge of industrial waste into bodies of water
- Discharge and dumping of waste materials and homemade products into bodies of water
- Oil spills

- Agricultural pollution via the runoff of pesticides and agrochemicals that are washed into water systems or otherwise infiltrate the groundwater
- Soil erosion sediments
- Natural disasters (floods, hurricanes, etc.) which result in the intermixing of water and dangerous substances on the land (Madaan, 2019)

Land. pollution results from the destruction, decline, and/or abuse in the quality of land resources with regard to their use and their ability to support life forms either directly or indirectly as a result of human activities. For example, land pollution occurs when waste products and trash are not disposed of correctly, and therefore introduces toxins and chemical into the land. Land pollution also occurs when humans dump chemical products (herbicides, pesticides, fertilizers or other consumer byproducts (correctly or otherwise) into the soil. The extraction and exploitation of mineral resources also contributes to the decline in the quality of the Earth's landmasses which has grave consequences for soil quality, plant life, and human health and well-being. Common contributors to land pollution are litter, deforestation, acid rain, solid waste, agricultural chemicals, construction and development, and mineral exploitation and depletion (Madaan, 2019).

In certain respects, *soil pollution* shares several commonalities with land pollution, such as contamination by chemicals or the exploitation of minerals through over-mining, but soil pollution also results from degradation from the clearance of vegetative cover and topsoil erosion which creates damage to the immediate environment. Consequently, the soil loses its valuable natural minerals and nutrients (soil degradation) which ultimately leads to soil pollution. In addition, agricultural farming activities, over-farming, over-grazing, and waste dumping on land are responsible for the consequences of desertification, loss of wildlife habitat, food contamination, and reduced crop yields. All this is happening at a time when the Earth's population is continuing to grow, a reduction in access to prime agricultural land and a reduced crop yield suggests serious and potentially fatal consequences in the future (Madaan, 2019). Famous American authors such as Steinbeck chronicled the gradual transformation of once viable indigenous ecosystems into mass agricultural enterprises, resulting in the rapid extinction of most native species, in exchange for the easy harvest of foodstuffs needed for the colonization of newly granted territory (Steinbeck, 2002).

Noise pollution differs from air, water, land, and soil pollution in that it does not accumulate in the environment. It is described as an undesirable sound(s) which causes discomfort for the ears, and it is calculated in decibels (dB) where levels above 100 dB may cause permanent hearing damage and loss. In current society, noise pollution has become a permanent and daily aspect due to construction and development, industrial manufacturing, concerts, sporting events, and transportation (airports, railroads, and traffic). Noise pollution not only adversely affects the human ears, but it can also impact human muscles due to the intensity of the vibrations; it is equally detrimental to non-humans and can even cause their death (Madaan, 2019). Noise pollution elevates stress levels and causes nervousness in both humans and non-humans, and may be harmful to neonates (Sustainable Baby Steps, 2010). It is also of note that artificial sounds (man-made noises), like addictive substances, can become addictive to the point where anxious individuals experience untenable malaise unless comforted by a constant stream of human information-like sounds—a sort of biophobic response to sounds in the air (BBC, 2019).

Thermal pollution is actually a wider aspect of water pollution which occurs when bodies of water are negatively impacted by the alteration (raising or lowering) of their temperatures. It is commonly associated with human activities and/or industrialization that changes the temperature in oceans, lakes, rivers, or ponds. Perhaps two of the greatest contributors to thermal pollution are power plants and industrial manufacturers who employ water as a cooling measure in their processes. Additionally, urban stormwater runoff from (concrete) buildings, parking lots, and roads is also a contributing factor in elevating the temperature in nearby bodies of water, and whenever there is some type of alternation (increase or decrease) in the temperature of a natural water resource, it tends to have dire consequences for the aquatic life as well as the local ecosystems (Madaan, 2019).

Industrial pollution is a triple threat as it releases industrial pollutants and wastes, such as air emissions, used water, landfill disposal, and the injection or burial (of drums) of toxic substances underground into the air, water, and land. Industrial pollution has the capacity to kill both humans and non-humans alike, degrade the quality of life as well as to cause imbalances in the ecosystem(s). Contributors to industrial pollution include, but are not limited to: power plants, heating, plants, steel mills, sewage treatment plants, and glass smelting plants (Madaan, 2019).

Light pollution is synonymous with *photo pollution* or *luminous pollution,* and occurs when there is an excessive use of artificial lights to a point where it causes a brightening of the skies which disrupts the activities and natural cycles of wildlife as well as being detrimental to human health. A nuisance is created whenever artificial lights are introduced where they are not intended to be, for example too much/excessive outdoor light or light shining in unguided/misguided directions (streetlights, porchlights, etc.) can disrupt individuals' sleep cycle. Light pollutants are described as *glare, light trespass* and *sky glow* with city lighting, advertising signage, billboards, and nighttime entertainment venues being some of the major contributors to light pollution particularly when they are run all night long (Madaan, 2019).

Radioactive pollution is the least likely type of pollution to occur; however, if or when it does happen, the impact is catastrophic to the environment as well as to human health. Due to its intensity, it is capably (and likely) to cause a massive human death toll, a myriad of serious if not fatal diseases (cancers, lung disease, organ damage or failure, birth defects) as well as significant destruction to plant and animal life. This type of pollution is capable of polluting the water, air, and soil, and it is likely to require many years for the environment to regain its natural capacity, but the sources of radioactive pollution are few although that is little consolation for the victims of such a disaster: nuclear power plant accidents, uranium mining operations, and the inappropriate disposal of nuclear material (Madaan, 2019).

Visual pollution tends to be more of a subjective classification of pollution which remains under-estimated and less-reported in contrast to the air, water and land categories. It can be described as anything that is deemed to be unattractive and/or potentially damaging to the surrounding natural landscape thereby rendering it "subjective." Some examples of visual pollution are: skyscrapers that block views of Nature, graffiti carved into trees, etched into rocks, or other types of natural landscapes, billboards, abandoned structures, etc. could also be considered as examples of environmental eyesores (Sustainable Baby Steps, 2010).

And so, there is a synopsis of ten ways that humans have developed that will effectively pollute, degrade and destroy the Earth; however, human beings can be especially creative when they have identified their personal, business, and/or financial goals and interests, so unfortunately the magnitude of human interference does not stop there if the concept of *instream barriers* and *altered waterflow* is taken into consideration. If a particular river, stream, lake or pond dare impede human expansion, development, or construction, it is simply moved or diverted someplace else where it does not hamper human interests and goals. Instream barriers include culverts, fords, dams, weirs and even pipes which are incorporated into infrastructures, such as roads and bridges, urban water supplies as well as for stormwater discharge into adjacent waterways. Altering the water flow simultaneously alters the habitat on which local species depend to live, migrate, and breed in addition to contributing to riverbank erosion and the disruption of riverbed habitats. New Zealand provides an excellent example of indigenous fish populations which are *diadromous* or are species that migrate between their adult environment and marine or lake environments are part of their natural life cycle. If instream structures are incorrectly installed or maintained, they are likely to restrict or completely prevent the migration of fish upstream and downstream which can ultimately modify the *hydrology* (water flow) of a waterway. Instream barriers that are not designed to allow the passage of fish, such as culverts that have been undercut, perched, sustain high velocity water flow, and/or lack wetted margins, do not permit some fish species, for example, tuna, to overcome these obstacles (Nukurangi, n.d.).

The detrimental effects posed by instream barriers are not just borne by the fish species that inhabit the nearby waters, but they also negative impact the modified channel form through erosion due to vegetation removal along banks as well as erosion caused by changes to the streamflow following the creation of a road crossing or similar type of barrier leading to their scouring and/or eventual breakdown. *Modified flow* or the realignment of the banks of streams can also change the *benthic* or bottom bed structure of the river or stream where substrates, such as gravel and rocks are replaced and/or covered over with sand resulting in habitat loss for native species. Further damage to banks and floodplains that vary water flow can initiate flash floods which will threaten the very stability of a riverbank, thereby increasing its vulnerability to collapse as well as damaging fish breeding and feeding grounds. The ongoing debate regarding climate change often challenges the concept of increasing water temperatures; however, even water flow affects water temperature so that a loss of water flow can create a fluctuation in temperature, and if left unshaded, can reach elevated temperatures unsuitable for local fish populations. Typically, fish are unable to tolerate temperatures over *two degrees higher* than their normal range, and when combined with decreased water clarity due to erosion and increased sediment deposits in a river will reduce visibility and impede their ability to locate food. A decrease in water flow can increase the level of nutrients dissolved in the water which may result in too high a concentration which, much like an elevated temperature, is not necessarily an optimal situation for the local fish population (Nukurangi, n.d.). Many scientists feel that the human species is to blame for the infamous *red tide effect*, which can arise from unusually high levels of nutrients favorable to the exponential blooming of these algal microorganisms, climate change, and similar anthropogenic changes in otherwise balanced ecosystems (Wei-

Haas, 2018). Some even hypothesize that one of the plagues of Egypt was due this very phenomenon.

So, the issue of altering, redirecting and/or damming the flow of rivers, streams, lakes, ponds, oceans, etc. for the convenience of humans is a complex one that extends far below the surface (pun intended). Oftentimes the far-reaching implications and long-term impacts are not identified for years if not decades, until one day a natural or man-made disaster occurs, or some infrastructure that has not been properly maintained collapses leaving environmental and/or ecological devastation in its wake. Perhaps the time has (finally) arrived to abandon the *dominion perspective* which assumes that this planet and everything on it or in it are for unlimited human control, consumption and disposal, and shift into a collaborative and mutually beneficial *biophilic perspective* where Nature is no longer *at the mercy of humans.* Nor is an attitude of tolerance toward non-human species sufficient (an ingenuine beneficence from a position of superiority), since this suggests that humans resent them, even if the entities subject themselves to human sovereignty. Nature does not need a sovereign to regulate what it limits naturally. Nature does not deserve our condemnation, nor does it need our forgiveness. What has Nature done to warrant such abuse, and who told humans they were in charge? Perhaps the same voice that told humans they were naked also persuaded them to take charge.

COLLABORATIVE EFFORTS: A GLIMMER OF HOPE?

An inter-species collaboration between both non-human and human species who occupy the Earth's same biosphere is more likely to result in maintaining Nature's balance and a much more peaceful co-existence among all life on this planet. Humans have much to learn from observing the behaviors of other animals who contribute so much in helping to maintain the delicate balance of life on Earth, as well to heal the damage humans have inflicted here. For example, scientists at the University of Washington have enlisted the aid of narwhals (which are a variety of whales who can dive much deeper than humans), to track the changing temperatures in Greenland's arctic waters. Certain bird species specialize in repurposing brightly colored plastic containers and bottle caps to attract mates; and while plastic has become a major ecological issue due to humans' indiscretions of disposing of trash, perhaps this will serve as a wakeup call in more urban areas where trash is being incorporated into the more traditional nesting materials. Another animal recycler is the octopus which is prized for its intelligence and ability to use tools; one particular species, the veined octopus, constructs its shelters from shells or other ocean debris which now includes glass jars found on the ocean's floor. Often marginalized as "pests," rats display a superior sense of smell—rivaling that of dogs—and now the African giant pouched variety has joined the defense forces of the anti-landmine organization known as APOPO due to their light weight and capability for locating environmentally hazardous land mines without setting them off. Bees are not only great dancers and vital plant pollinators, but also assist humans in detecting toxic chemicals in the environment which could be related to potential chemical warfare attacks (Phair, 2018). Perhaps the unique capabilities, adaptability, and wisdom of non-human species such as these, will serve as an inspiration for humans in their attempts to survive despite the degradation and destruction of the Earth's biosphere due to human interference and mismanagement. Granted these examples are not indicative of a 50:50 type of partnership between non-humans and humans; however, even the realization that other species display talents and abilities that far exceed the capabilities and limitations of the human condition is (seemingly) a step in the right direction.

SHIFTING INTO NATURE'S PERSPECTIVE VIA A BIOPHILIC LENS

These authors suggest that the time has come for humans to abandon their myopic mindset—an unjustified belief that humans are the most fit species on Earth, destined to rule over their world. This is clearly not the case, given the number of environmental and ecological crises that are being endured by *all* species that occupy this biosphere, due to the actions of just one arrogant species. The time has come to shift into the collaborative mindset of biophilia where Nature is most certainly not the enemy . . .

Remember:

Within the earth's biosphere, all species both non-human and human are inextricably linked and innately motivated to consistently focus on and to interact with other species. When these motivations to focus on and interact with other species demonstrate the characteristics of awe, reverence, respect, and/or empathy, these inter-species interests and interactions can be mutually beneficial psychologically, biologically, emotionally, and spiritually in ways that encourage the species' survival, evolution, development, and ability to flourish. Conversely, when these inter-species interests and interactions exhibit characteristics that are indicative of egocentrism, self-serving biases, devaluation, and domination of one species by or over another species, the outcomes can be catastrophic not only for that specific interspecies exchange but also across the entire biosphere resulting in the decimation, destruction, and/or extinction of both known and yet-to-be discovered species. (O'Grady, 2016)

CONCLUSION

This chapter has highlighted the problem—the widespread belief that the human species was predestined to reign over the natural world. First, human decisions are extremely self-centric and myopic, leading to unsustainable ecosystems that require deliberate extermination of flourishing populations to rebalance the system. Second, wilderness and wildlife management practices suggest a long history of wildlife exploitation under the auspices of nurture and protection, from ancient kingdoms to modern enterprises in which the welfare of wildlife is secondary to the welfare of its management. The expert solutions attempted (the introduction of non-native species to rectify unnatural transformation of the ecosystem for human affairs) evince a complete lack of understanding of the natural volume of time required to formulate and test a complex living system. In Nature, there is no equivalent for the human idea of manufactured "products" which have no purposes of their own. In the wilderness, there is no inevitable piling of "waste products" that render entire regions uninhabitable via pollutants in every biome. Some surprising side effects of human pollution have been the repurposing of human waste for natural habitats by non-human species, when faced with few other choices for a habitat. Perhaps rationalism is not capable of adaptation after all (because it puts its own ambitions before the basic needs of living things; Bonjour, 1997). Perhaps what is needed is to hand the reins over to those who have proven their stewardship capabilities for eons (this abundant flora and fauna), not add more executive control over things which only wish to be free to choose their own destinies (Wilson, 2016). The very idea of humans "nurturing Nature" seems to be oxymoronic. At the very least, it seems an ironic point of view to see ourselves as Nature's guardians, seeing how carelessly the human species has treated its world thus far.

REFERENCES

BBC. (2019). *The noisy planet*. Retrieved from http://www.bbc.co.uk/worldservice/specials/1643_noisyplanet/page3.shtml

Bonjour, L. (1997). *In defense of pure reason: A rationalist account of a priori justification*. Cambridge, UK: Cambridge University Press. doi:10.1017/CBO9780511625176

Burns, R. (1785). *To a mouse, on turning her up in her nest with the plough*. Retrieved from http://www.robertburns.org/

Chinese Grass Carp. (2019). *UF/IFAS University of Florida, Plant Management in Florida Waters – An Integrated Approach*. Retrieved from http://plants.ifas.ufl.edu/manage/control-methods/biological-control/chinese-grass-carp/

Christensen, D. (2018, September 24). Invasive species threaten Florida. *Orlando Sentinel*. Retrieved from https://www.orlandosentinel.com/news/os-invasive-species-florida-pictures-photogallery.html

Darwin, C. R. (1876). *The origin of species by means of natural selection, or the preservation of favoured races in the struggle for life* (6th ed.). London, UK: John Murray.

Dolbeer, R. A., Chipman, R. B., Gosser, A. L., & Barras, S. C. (2003, May). Does shooting alter flight patterns of gulls: case study at John F. Kennedy International Airport. In *Proceedings of the International Bird Strike Committee Meeting* (Vol. 2, pp. 547-564). London, UK: IBSC.

Gaiman, N., & Pratchett, T. (2019). *Good omens: The nice and accurate prophecies of Agnes Nutter, witch*. New York, NY: HarperCollins.

Gallagher, B., Hogg, S., Marmion, G., & Trickelbank, A. (Producers). (2009). Lark Rise to Candleford. United Kingdom: BBC One.

Hodson, G., MacInnis, C. C., & Costello, K. (2014). (Over)valuing "humanness" as an aggravator of intergroup prejudices and discrimination. In P. G. Bain, J. Vaes, & J. P. Leyens (Eds.), *Humanness and Dehumanization* (pp. 86–110). New York, NY: Psychology Press.

Hunt, P. (2007). *Reading livers through reading literature: Hepatoscopy and haruspicy in Iliad 20:469 ff & 24:212 ff, Aeneid 4:60 ff & 10.175 ff, Cicero and Pliny on divination, among others*. Retrieved from https://web.stanford.edu/dept/archaeology/cgi-bin/archaeolog/?p=145https://en.wikipedia.org/wiki/Augury

Madaan, S. (2019). *What is pollution?* Retrieved from https://www.eartheclipse.com/pollution/different-types-of-pollution.html

MarineBio. (2019). *Alien species.* Retrieved from https://marinebio.org/conservation/marine-conservation-biology/biodiversity/alien-species/

Moses, D., & Stone, D. (2013). *Colonialism and genocide*. Hoboken, NJ: Taylor and Francis. doi:10.4324/9781315878713

Nukurangi, T. (n.d.). *Instream barriers and altered water flow*. Retrieved from https://www.niwa.co.nz/our-science/freshwater/tools/kaitlaki_tools/impacts/in-stream-barriers

O'Grady, M. A. (2016). *The theory of biophilia* (Reg. No. TXu 2-032-922). United States Copyright Office. Retrieved from https://www.copyright.gov/

Phair, C. (2018). Animals that help save the planet. *Pachamama Alliance.* Retrieved from https://blog.pachamama.org/anaimals-that-save-the-plant

Quinn, J. (2015, June 8). The laws of physics says that two objects cannot occupy the same space at the same time. Yet, electromagnetic waves can. How does physics apply to electromagnetic waves? [Blog comment]. Retrieved from https://www.quora.com/The-laws-of-physics-says-that-two-objects-cannot-occupy-the-same-space-at-the-same-time-Yet-electromagnetic-waves-can-How-does-physics-apply-to-electromagnetic-waves

Slezak, M. (2015, March 19). Cane toad has surprise effect on Australian ecosystem. *New Scientist*. Retrieved from https://www.newscientist.com/article/dn27199-cane-toad-has-surprise-effect-on-australian-ecosystem/

Smithee, A. (2019). *Five facts about rites of passage in native American cultures*. Retrieved from https://www.ranker.com/list/facts-about-native-american-rites-of-passage/alan-smithee

Steinbeck, J. (2002). *East of Eden*. New York: Penguin Books.

Stokes, M. (2011). An introduction to managing wild game populations. *Nature Education Knowledge, 2*(11), 4. Retrieved from https://www.nature.com/scitable/knowledge/library/an-introduction-to-managing-wild-game-populations-235686790

Sustainable Baby Steps. (2010). *Seven kinds of environmental pollution.* Retrieved from http://www.sustainablebabysteps.com/kinds-of-environmental-pollution.html

Wei-Haas, M. (2018, August 8). Red tide is devastating Florida's sea life. Are humans to blame? *National Geographic*. Retrieved from https://www.nationalgeographic.com/environment/2018/08/news-longest-red-tide-wildlife-deaths-marine-life-toxins/

Wilson, E. O. (2016). *Half-earth: Our planet's fight for life*. New York, NY: Liveright Publishing.

World Health Organization. (2019). *Sex ratio.* Retrieved from http://www.searo.who.int/entity/health_situation_trends/data/chi/sex-ratio/en/

Section 3
The Cure for Societal Indifference

This introduction to the third and final book section defines the purpose of this third part. The four chapters included in this section include "Losing Everything to Save the World," "First Love: How Conflict Resolution Can Cure Indifference," "How Reverence for a Flower Can Save a Planet," and "Beyond First Love: No Greater Love." This introduction simply outlines the themes that the third set of four chapters will cover, serves as one of the dividers of the material into three logical chunks, and helps outline how they fit into the organization of the book's three objectives.

The theme of this third and final section is about faith—faith in the capacity to rise above these petty differences and this curious indifference toward the living planet that all living things call Home. The authors believe there has always been a way to save this world from Mankind's shortsightedness. That way is to do *that one thing* men and women of valor have never done before . . . let the planet's welfare be unusually dearer than their own.

In the closing chapters of this treatment of biophilia, a ray of hope will be discussed which the authors hope the reader will embrace and license no delay in deliberating its potential. It is not an easy cure; few of this species, extant today, who selflessly commit their common future to its success, will ever know for certain whether the patient they gave their all to resuscitate made it through the night. But they will know they gave their all. As President Kennedy once said to an awe-inspired audience in Texas, but rephrased for this topic, "We choose to save the planet Earth. We choose to save the planet Earth in this decade and do the other things, not because they are easy, but because they are hard, because that goal will serve to organize and measure the best of our energies and skills, because that challenge is one that we are willing to accept, one we are unwilling to postpone, and one which we intend to win, and the others, too" (NASA, 1962, para. 16).

It is time to choose to do a thing never done before, not because it is possible, but because there is still time to do the impossible. It probably isn't reasonable, and most (if not all) may, in the final analysis, lose what rights they have so long revered by gifting them back to something they revere more; but it is surely an act of conscience, and most certainly one of uncommon dignity and purpose.

REFERENCES

NASA. (1962). *JFK Rice moon speech.* Retrieved from http://er.jsc.nasa.gov/seh/ricetalk.htm

Chapter 9
Losing Everything to Save the World

ABSTRACT

In the authors' research, it was discovered that when people lose a little of their many possessions due to a disaster, they become unexpectedly upset as though they were cheated out of their share. This happens even if those who behave indignantly are otherwise nice people. They grumble over having to make repairs instead of being grateful to have survived. But curiously, when they lose absolutely everything to a disaster (even though one would expect them to be even more upset), they become surprisingly grateful instead (almost relieved). The authors propose that the unexpected indignation at losing a little is really just the habituation of gratitude coming out at last, because long ago a little was all humans had to lose. But as humans have too much to lose these days, it is hard to get back to the sense of losing everything. Extreme loss is required to rediscover extreme relief, and this requires letting natural outcomes determine when and where one loses everything—not by greater control of outcomes, but by broadening the idea of faith.

INTRODUCTION

In this chapter about losing everything, the authors will show that imagining an outcome is not all that safe. This chapter will talk about things humans cannot control, and how a complete absence of human control often results in a more complete gratitude for having survived. The authors call this phenomenon the *Hurricane Survivor Effect*. This chapter will discuss extreme loss and extreme

DOI: 10.4018/978-1-7998-4408-2.ch009

relief, and how there is more to the idea of faith in the future than simply enslaving that future so that it won't disappoint people. This chapter will ask the reader to imagine the abolition of differential treatment of Earth's species, and challenge the idea of courage, so that it includes taking unwise risks for something so beloved the idea of risk itself is suddenly attractive. Lastly, this chapter will challenge humanity's dearest assumptions to see if buried there in that biased certainty is the very antidote that might save the non-human world from an early death. Perhaps, the way to save a dying planet is to lose the very thing the human species is most afraid to lose (Meinecke, 2017).

FORCE MAJEURE

Humans are obsessed with dominating outcomes. Humans do not like uncertainty—to them it is a sin, a mental deficit, or a crime not to predict and control one's personal outcomes (even if they are plainly outside one's control). And if there is any way to influence Fate, the human species will try to influence or modify Fate to favor its species (usually by offering up sacrifices to Chance; Meinecke, 2018a, 2018b). In fact, this obsession with favor does not stop at a species boundary, but goes on and on aggressively within the species too—so that even similar people who share the same region of the world (or live next door to each other) compete for political favor against their neighbors. Humans must know the future, and once humans know it, they must change that future, so it benefits them and nobody else. This way humans gain a "pawn-up" advantage, and in the end game, that little advantage helps humans out-survive their neighbors.

In spite of this need to know and control the future, humans cannot control everything. From the ancient practice of *haruspicy* (the reading of animal entrails) to the medical practice of autopsy (the inspection of human entrails), from the bizarre practice of augury (bird signs) to the modern science of meteorology (atmospheric signs), humans cannot always guess the future. The *scientific method* itself is founded on the idea that what is natural cannot be guessed or controlled (scientists call it randomness). So, though it may be crucial to social fitness to predict the future, there is a random gap in what humans can control; humans even have phrases for it. In law they call it Force Majeure; in the psychology of religion they call it God of the gaps; in the commercial need for indemnification from blame, they call it an Act of God (Gorsuch & Smith, 1983).

THE HURRICANE SURVIVOR EFFECT

As a consequence of the incapacity to know the future with any degree of certainty, natural disasters exact a toll on human civilization because humans build upon Nature as though Nature would never injure them—simply because humans are special. The reality of course is that humans are not special, and a tempest will sweep away thousands of human beings as non-prejudicially as thousands of cattle. From floods to hurricanes to famine, natural disasters exact tribute from a faithless species (faithless in the sense of disbelieving that Chance dictates their future, not their leaders or their beliefs)—which stubbornly builds settlements and pens animals up where natural forces (including diseases) find them easy prey. Then humans get upset because disasters strike, even if humans could simply become nomadic and avoid them entirely like their ancestors did. A tent does not crush its occupants the way a brick building does, yet humans do not allow people to live in tents and make their building standards stiffer instead. There is therefore an inescapable idea of acceptable losses with respect to adaptations humans would rather not make. Some of their friends will lose what is dear to them, and humans hope it isn't them when it occurs.

There is a more significant finding though, in this concept of natural losses (caused by God or Nature) versus unnatural losses (caused by individuals or society), and it holds great promise in changing their attitudes toward Life and each other. In their research, the authors discovered that when people lose a little of their many possessions due to a disaster, they become unexpectedly upset as though they were cheated out of their share (even if they are otherwise nice people). They grumble over having to make repairs instead of being grateful to have survived. But when they lose everything to a disaster—even though you would expect them to be *even more* upset—instead they become surprisingly humble, and just grateful to be alive. Now that they have lost everything, they realize that "everything" has always consisted of one another—their loved ones, not their stuff. They suddenly appreciate their wives and children and neighbors, now that all of their material goods have been lost. Having no one to blame for their misfortune, they become thankful instead of upset at this sudden turn in their fortune. It does not seem to be a character quality that disasters reveal, so much as the accidental ability to finally see what really mattered to them.

EXTREME LOSS: LOSING EVERYTHING

Ken Druck (2017) is an expert on loss and how to help others manage to get through it. After surviving the personal tragedy of losing his daughter in 1996, helping others became his reason for going on. Ken had lost his sense of purpose and his motivation for living; he watched as others endured seemingly insurmountable losses as a result of natural and man-made disasters—such as hurricanes, earthquakes, unspeakable acts of violence, illness, addiction, etc. Just as Life as he knew it had ended, Life as they knew it had ended. Just as Grief had become his day-to-day companion instead of his daughter, Grief had become their new partner. He knew what it felt like, what it *meant* to wake up each day, knowing this problem would never be fixed—that his daughter would never come back. Life as he knew it, Life as they knew it, had ended; a new reality had begun to set in, a reality consisting of an affective inventory of what endearing mementos of his prior purpose for being, remained to live for and aim for.

He noted that many non-grieving individuals try to help the grieving cope, in their roles as first responders, law enforcement, firefighters, neighbors, federal agencies, and non-profits. Maybe mobilizing these resources and taking steps to help in any way they can, could alleviate some of the heartache. These range from local to global efforts supplying food, water, clothing, shelter, financial aid, compassionate thoughts, prayers . . . and even hugs.

Druck (2017) noted that, in addition to these generous donations of tangible items, emotional and psychological support provide significant healing effects to those who have endured life and living losses. Some examples include:

- Be empathetic and patient with individuals who have suffered a catastrophic loss, even if it seems they do not want help at this stage of the grieving process, and no matter how much effort it requires
- Suspend all judgments, criticism, opinions, and impatience by offering only kindness to individuals who may be feeling guilt, fear, despair, and sorrow when their world has been shaken to its very core and their sense of purpose compromised
- Exhibit compassion and understanding for individuals whose future for themselves and their families may have been completely obliterated or at best remains in question
- Initiate and mobilize support services by bringing survivors together to support one another in addition to providing funding and access to vital resources

- Pitch in and help by volunteering wherever and whenever assistance is needed
- Provide life support by helping individuals in the local community who are living with losses as a result of natural disasters, homelessness, physical and/or psychological illness or disorders, trauma, and similar, whether they are struggling to stay afloat or mired in the depths of despair

The resilience of the human spirit to rise above a sense of overwhelming loss of loved ones, homes, workplaces, possessions, etc. is evidenced by an individual's ability to help and support others while personally recovering from a catastrophic loss. (Often, this is why survivors come together—who better to provide compassion and empathy, as well as engage in mutually beneficial activities, than someone who has "been there" themselves? (Druck, 2017).

Above and beyond humans helping humans cope with grief, the subfield of *trans-species psychology* (Bradshaw & Watkins, 2006) expands the idea of bidirectional benefit, as posited in *The Theory of Biophilia* (O'Grady, 2016). Both feelings (mutually beneficial empathy) and actions (mutually beneficial activities) may be employed, not only among conspecifics, but between species in need of meaning and the ubiquitous search for a purpose. Evidence of such feelings and behaviors is demonstrated by *Psychology Today*'s documentary entitled, *Odd Couples: Compassion Doesn't Know Species Lines* (Bekoff, 2012).

ALTRUISM MAY EXPLAIN UNEXPECTED JOY AFTER EXTREME LOSS

From the topic of irreconcilable grief, the opportunity arises to discuss irreconcilable joy. Just as tragedy strikes with no warning or cause, often redemption strikes with no foreshadow or delayed explanation. Random acts of Nature's kindness, also called altruism, seem to occur as inexplicably often as random acts of Nature's wrath. Tor Norretranders (2014) suggests that the time has arrived for the concept of *altruism* to be "retired," but not because the phenomenon has become obsolete where humans are clinging to that ongoing debate that pertains to their egocentric tendencies, quite the opposite. Norretranders suggests that the human appreciation for the significance of the bonds existing between humans and non-humans is actually increasing due to the acquisition of a better comprehension of human and non-human societies. The issue with the concept of altruism involves the underlying belief that a conflict of interest exists between one's ability to help oneself while helping others. Whose interest is first? Many believe that both humans and non-humans are inherently selfish and egocentric, and that nothing is done without hope of compensation. The concept of altruism fills in a gap in this absolute belief in *social exchange theory* (Tulane University, 2020), to help explain those behaviors indicative of genuine kindness with no hope of compensation.

To explain, social exchange theory holds that all human relationships—even romantic ones—are driven by behavioral economics and logic, rather than out of simple love for one another, despite the absence of a sustainable solution to Life's problems (John 15:13, KJV; Tulane University, 2020). As a result of this line of reasoning, if an individual loves another individual no matter what, adherents of social exchange theory hold that something must be wrong with that individual's attachment style, cost-benefit analytical ability, self-image, or logic. But some hold that an indefeasible reverence for all Life, not the economics underlying the guaranteed survival of their genes, is sufficient reward for loving and helping others (Meinecke, 2017). Parental investment theory and evolutionary theory equally exhibit gaps in their ability to explain why some acts of love seem to defy explanation--absolute love out of nowhere, and with no hope of requital. Perhaps the most memorable example was recorded long ago: "He was taken from prison and from judgment: and who shall declare his generation? For he was cut off out of the land of the living: for the transgression of my people was he stricken" (Isaiah 53:8). No greater love has any man, it seems, than to do an inexplicable act of mercy for the most undeserving thing.

On the other hand, it may actually be the case that humans were once born kind with no hope of compensation and have lately become selfish, rather than born selfish and finally become kind. Keltner (2009) posited that there is strong evidence in favor of this contrary explanation for otherwise fickle acts of kindness. He explained that the hope of social reciprocity and financial health are not the natural condition nor motivation of the human species (Keltner, 2009). If so, then by their unusual prosperity humans have accidentally impoverished that once charitable condition of pricelessness, innocence, and biological trust—yet not pruned entirely away such a Nature. Children, for example, display an uncanny mercy for Life of every kind; then grown-ups force them to "grow up." When they do, they end up as ungrateful and self-praising as every other grown-up who has survived the brief spring of childhood, only to endure the long winter of discontent called adulthood

(Meinecke, 2018a, 2018b). Therefore, to these authors, the starting point for all social interaction is probably the hope of trust, not the hope of compensation (a mutual expectation of unconditional regard instead of collective prosperity). It is plainly an illusion to believe that one individual or group can be blithely happy while frequently surrounded by wailing conspecifics, whose constant reminders of one's unusual good fortune go neglected. Some say that everyone experiences grief at least once (Manning, 2011). It is likely that humans begin sympathetic toward all biological things (since humans are of the same biological persuasion), but the outrageous tares of life make it too hard to feel sympathy for everyone (and survive one's own tragedies too). So, humans limit their pity and affective bonds toward others (to limit their chances of collapsing under their own loneliness and grief). It is unlikely that those with too little grief do not comprehend the infirmities of those with too much, or because some were born with too little pity and others were born with too much (Zimbardo, 2007).

SCHADENFREUDE MAY EXPLAIN JOY AFTER ANOTHER'S LOSS

An exception to embracing this type of bond is called *Schadenfreude*, which can be described as the distinctive, vicarious type of pleasure that individuals derive from witnessing the excess misfortune of others when they cannot enjoy excess good fortune directly (Wang, Lilenfeld, & Rochat, 2018). Research that has been conducted over the past 30-years has indicated that Schadenfreude is multifaceted in nature, and is anchored in humans' concerns for social justice, social identity, and self-evaluation—or the process of a critical evaluation of an individual's performance based upon specific standards that were previously agreed upon. The more recent motivational model of Schadenfreude proposed by Wang, Lilenfeld, and Rochat in 2018 suggests that it is comprised of three separate yet interrelated subcategories (aggression, rivalry, and justice) which indicate different developmental trajectories and personality traits. This motivational model also purports that *dehumanization* assumes a vital role in eliciting Schadenfreude as well as integrating its various components.

To play devil's advocate for a moment, in the field of *experimental economics,* from a psychological perspective it is logical that if helping others provides "warm and fuzzy feelings" to an individual, is it not rewarding and in that individual's best interest to help others? Wouldn't that individual simultaneously be helping himself or herself, even if that may not be his or her primary motivation for helping others (Norretranders, 2014)? Although according to a study recently published in the *Journal of Experimental Psychology: General,* (Cameron et al., 2019) suggests that individuals attempt to avoid feeling empathy for other humans because of the mental and emotional effort it requires. Researchers recruited more than 1,200 subjects over the course of 11 experiments which indicated that subjects chose from the "empathy" deck of photos only 35% of the time in contrast to the choosing from the physical characteristics of the individual pictured deck the remainder of the time. The majority of the subjects reported afterward that not only did feeling empathy for the individual pictured required more effort, but that they also felt less successful at feeling empathy than they did for identifying physical characteristics.

Is altruism just another measurable trait? This is the troubling part of how often the legitimate sciences view folk-science, faith-based explanations, and ethological motives (such as unjustified charity and unreciprocated mercy). It is hard to conceive of doing a thing one would never do. Some would never do good simply for the sake of doing a good thing; they think only of the outcome, never the struggle to do what is good as an end in itself. Altruism does not seem to care about that (who does good for whom or why). It does not seek credit; it does not attribute blame; it is more like a kind of faith than a given belief or practice (Meinecke, 2020b). Their disbelief, then, prevents them from hearing that it exists in others. This too has been recorded since ancient times, whether acts of extreme kindness, or acts that take extreme faith:

Peter got out of the boat and walked on the water and came to Jesus. But when he saw the wind, he was afraid, and beginning to sink he cried out, "Lord, save me." Jesus immediately reached out his hand and took hold of him, saying to him, "O you of little faith, why did you doubt?" (Matthew 14:29-31)

And then there is the psychological or scientific explanation, given that the idea of universal mercy may be viewed under a scientific lens or one of conspecific trust. It could be as simple as a visceral appreciation of the energy available and the competence to achieve it, rather than reluctance to exert more effort than less. The study of performance psychology and the study of rumination have discovered that, if there is an unusual benefit resulting from imagining what will occur in order to avoid it (versus trusting in whatever happens), it is logical to assume there is also an unusual cost resulting from imagining what will occur in order to prosper from it (Davey, 2011; Lyubomirsky & Tkach, 2003). So, it requires more energy to contemplate doing good, than simply to do good, and more energy to practice a mercy based on merit, than to practice random acts of mercy (Meinecke, 2017; Meinecke, 2018a, 2018b). If prosperity is not one of the conditions for mercy, mercy may be distributed more equally.

PAIR-BONDS MAY EXPLAIN INTERSPECIES JOY

Sometimes, in the course of non-human events, the need arises for a less than perfect pair to form a more transitory union, to outmaneuver those unassailable forces which by sheer numbers of able-bodied conscripts pair and predominate for centuries. Two unlikely castaways may come together for comfort—for example, a traumatized man back from a foreign war and a traumatized dog left to die by its former owners. Pets for Vets is one example of such an unusual mercy for both species resulting in trans-species joy (Pets for Vets, 2020).

Why, though, is mercy so hard to come by? Why does the erstwhile tendency to exhibit *a more random mercy* go "aft agley?" (Burns, 1785). Perhaps this unnatural mercy for human concepts is greater than a natural mercy for living things? Perhaps humans mistake group-bonds (a conceptual union) for pair-bonds (a physiological union)? For it is difficult to pair-bond in the physical sense if one of you is conceptual by nature; but it is quite easy to pair-bond if both of you are physical by nature. Yet isn't it curious that two or more of anything (physical or not) is always conceptual by nature (Meinecke, 2017). So if humans have trouble sharing bonds with animals, perhaps it is because humans are trying to bond with the idea of an animal rather than just that one animal companion? Much evidence suggests humans readily pair-bond with their pets—and humans rarely have more than a few pets. Little evidence suggests that humans pair-bond with their livestock—and humans rarely have just one of those livestock. Humans use far more than one animal companion to serve as their food or labor or clothing supply—perhaps because it is hard for one soul to form an intimate bond with more than one soul at a time, thus justifying the view of many companion animals as supplies.

In any case, the existence of the intimate bond shared by humans and non-humans has been demonstrated through behavioral science and neurobiology via "synchronizing" phenomena, such as:

- mimicry (the action of imitating someone or something)
- mirroring (matching or simulating a behavior)
- emotional contagion (expressing and/or feeling emotions similar to those emotions demonstrated or felt by others; Nakahashi & Ohtsuki, 2018)
- empathy (comprehending and sharing others' feelings)
- compassion (concern for others' misfortune and/or suffering)
- sympathy (sharing a common feeling and understanding)
- other types of prosocial behaviors (voluntary behaviors designed to help others which is not exhibited only by humans which can include making donations to charity, rescuing a stranded motorist, one primate grooming another, or an adult dog playing more gently with puppies than when playing with other adult dogs (Prosocial Behavior, 2019).

REQUIEM FOR A TRANS-SPECIES PSYCHOLOGY

In 2006, Bradshaw and Watkins published their article purporting the theory and praxis of *Trans-species Psychology* which states that humans and non-human animals share common factors in cognition or thinking as well as in emotions or feelings. According to Bradshaw, her "science of sentience" was based upon data and information stemming from neuroscience, psychology, and ethology that dates back to Darwin's evolutionary biology in the 1800s that indicated the trans-species conservation of the brain and the mind (Bradshaw & Watkins, 2006). Therefore, humans and other animals share a common ability to think, feel, experience life, and themselves as well as in some cases to experience and to demonstrate empathy, compassion, self-awareness, consciousness, trauma, suffering, mourning, and complex communication capabilities. This trans-species paradigm has challenged humans to re-think virtually every aspect of modern culture as well as to reconstruct human identity as a much more egalitarian relationship with all non-human species (Bradshaw, 2010a). A combination of trans-species psychology and science upended Aristotle's *scale naturae* paradigm which ordered nature from lower species to higher species with humans at the apex (Marino, 2011). Trans-species science also rectifies the inconsistency generated by the practice of *unidirectional inference* which is the accepted practice of making inferences regarding humans from non-humans but not regarding non-humans from humans; engaging in *anthropomorphism* was considered to be unscientific. Unidirectional inference has been categorized as being inconsistent with scientific theory and research which supports the concept that the mental states of non-human species can indeed be inferred from humans just as genetic, physiological and morphological traits can be inferred, i.e. *bidirectional inference.*

Trans-species psychology has profound implications for comprehending the relationships as well as the interdependence between the human and non-human species. Bradshaw and Watkins (2006) quoted C.G. Jung's articulation of the negative impact humans' detachment with Nature has on the human psyche:

Man feels himself isolated in the cosmos, because he is no longer involved in nature, and has lost his emotional 'unconscious identity' with natural phenomena . . . No voices now speak to man from stones, plants, and animals, nor does he speak to them believing they can hear. His contact with nature has gone, and with it has gone profound emotional energy that his symbolic connection supplied. (p. 6)

This interdependence between human and non-human species is based upon the following principles:

First, trans-species psychology recognition of commonalities between human and non-human species "re-embeds" humans back into Nature's continuum where relationships across species are horizontal in representation rather than placed on a hierarchical or vertical gradient of inequality. Second, previous research examining human-to-human violence, aggression, and domination, such as slavery, genocide, etc. indicates that being in the role of oppressor or the oppressed creates a pathogenic psychological state whereby continued mistreatment and domination of non-human species who embody minds and capabilities similar to humans, causes suffering for all (Bradshaw & Watkins, 2006). Regarding an individual as less than human involves a part of the brain that is distinct from the part of the brain that is involved in simply disliking him or her according to a research study (Bruneau, Jacoby, Kteily, & Saxe, 2018) recently published in the *Journal of Experimental Psychology: General.* Researchers employed fMRI to scan the brains of 24 subjects as they were questioned regarding 10 different groups which included Europeans, Americans, Roma, Muslims, homeless populations, puppies, and rats. Certain questions were designed to assess like or dislike, in other words how *cold* or *warm* subjects felt toward that group, while other questions assessed dehumanization or where subjects felt that the group belonged on an *ascent of man* scale depicting evolutionary stages. This study revealed that regions in the left inferior parietal cortex and left inferior frontal cortex of the brain were associated with subjects' ratings indicating dehumanization. In contrast, a region in the posterior cingulate cortex of the brain was associated with a like or a dislike response from the research subjects.

The reference of trans-species psychology as being a "science of sentience in service" seeks to expand the comprehension of the human-non-human bond by eliminating the hierarchy between species that is anchored in differences, and replace it with mutual respect and parity to ensure that the differences between the species are not denied, but rather viewed with an appreciation for diversity (Bradshaw, 2006). Therefore, non-human species are extended the right to individual agency as well as species-specific cultural determination which results in major ethical implications across many fields including *ecopsychology* which uses non-human species in service to humans as well as keeping non-human species in captivity as a means for benefiting human health and well-being, such as in Animal Assisted Therapy or AAT which can involve dolphins, horses, dogs, etc. (Bradshaw, 2010b; Savage-Rumbaugh, Wamba, Wamba, & Wamba, 2007). Trans-species psychology suggests that any type of animal-assisted therapy program be made to adhere to the same ethical standards which are granted to human children and others who are not able to grant their formal consent as a means of avoiding exploitation by their participation in the program. In addition, TSP suggests that such standards should be developed by independent experts in the field of animal welfare rather than by members of the AAT industry itself. The common traits in cognition that are displayed by humans and non-humans support the fact that the exploitation of non-human species has the same potential for causing trauma in them as it does in human species (Borchers & Bradshaw, 2008).

Trans-species culture embodies seeking ways to peacefully co-exist with non-human animals without violence toward them or destroying their bodies, families, self- determination, or habitats. (p. 41)

Further discussions in the field of trans-species psychology conceptualize making a shift from the traditional perspective of humans' attempts at non-human species preservation and wildlife management to adopting a lens of embracing a multi-species culture and letting animals determine their own destiny. The core values of trans-species psychology discourage such inhumane practices as systematic extermination or *culling,* sustainable harvesting, captive breeding programs, etc. since they create emotional trauma, destroy family bonds, and ultimately cause a breakdown in multi-species culture. The realization that non-humans experience complex emotions and are susceptible to the effects of stress, violence and trauma just as their human counterparts, makes these practices not only disturbing but downright abhorrent (Bradshaw, 2009).

CIVILIZED HUMANITY AS THE DEVOLUTION OF MERCY

If the idea of chivalry evolved from war and has now devolved into an absence of chivalry in war, perhaps mercy is also mistaken to have evolved from cruelty and is now devolving back toward an absence of mercy from industry? (Gershon, 2019). Perhaps the idea of an intercity park is mistaken for an abundant mercy, whereas these lovely oases in the middle of vast deserts of stone and tar suggest that mercy is far more scarce than abundant. Abundance does not make what is rare more precious but more common.

Many individuals who serve as advocates for Nature and the significance of Nature in everyday life, focus on what version of Nature happens to be nearby which can be parks, gardens, walking trails, family pets, and/or birds attracted to feeders that have been erected in urban settings. However, domestic Nature is only part of the equation where the other part is wild Nature which has impacted the evolution of all species, and so this innate need for wildness remains a part of the mind, body, and spirit. That unpredictable wildness which can be described as being untamed, unmanaged, unfettered, unencumbered, unmediated, and self-organizing, in essence, the embodiment of all those traits that the human species has lost which often creates conflicting feelings of fear yet attraction, nurturing yet strength.

Perhaps the human species has actually *devolved* and become "soft" as a result of living in urban settings due to the reduction in the amount of wild lands and wild Nature that remain unspoiled by human interference. Therefore, subsequent generations of people have much fewer opportunities to interact with the "wild side" which results in a baseline of behaviors and interactions with diminished wild landscapes that becomes "normal" or what is considered to be "wild Nature" (Kahn & Hasbach, 2013). Kahn (1999, 2011) coined the term, *environmental generational amnesia* to describe how cohorts of subsequent generations each create their conception of what constitutes "environmentally normal" which is anchored in the natural world they encountered as children. In effect, nature and nurture are each amnestic with respect to the other's need (Meinecke, 2020a). So, as the amount of environmental degradation and loss of wilderness increases, this reduced level or condition is deemed to be nondegraded and normal for them; however, it is not as simple as the human species adapting to this level of diminished Nature without knowing about it, and doing fine, but rather it is a situation where the humans species adapts to this level of diminished Nature without knowing it, and ceases to flourish as individuals as well as a species. Kahn's analogy for this situation is being born in a prison, and having developed no concept of what world lay beyond the prison's walls. This has some evidence as well—there are some grown men who have never seen the surface of the Earth because everywhere they have ever walked was paved over (R. Delarose, JD, personal communication, September 19, 2016) Most of those native prison inhabitants could live full lives in biological years, but not as measured by species' standards which had been shaped by eons of evolution in Nature, so essentially as the wildness is lost, humans as a species are diminished.

This chapter might expand on this idea, since it applies not only to Nature but to knowledge itself and goes well with the other chapters on why it is so vital to save the planet, not just to record it for posterity. The ability to describe a thing without visiting it diminishes the desire to go and visit it for oneself; the taxonomy of a species diminishes the urgency to preserve the living members of that species (since how can someone be curious about what they already know?). Facts have this odd side effect (to know without going, rather than to go and see). To use Kahn's analogy (1999, 2011), facts are prisons; taxonomy is a prison. One is only as safe in one's finite cell as one refuses to look outside that cell for the infinite. It is not possible to know everything and also be amazed by what you know. But learn just one thing, and suddenly *everything you don't know* is amazing. Ergo, to truly revere Life, learn to revere just one thing, and then use your imagination to wonder how amazing the rest must be. The authors will go into this more in chapter 11 (*How Reverence for a Flower Can Save a Planet*).

Civilizations can often be identified by their technological trappings where people who reside in urban settings abandon their creature comforts of home to venture out of the safety of their confines to experience Nature by camping in the woods if only for one night. A decision such as this permits them to experience a *primal encounter* and sensorial access to their surroundings including the night sky which is often made invisible to city dwellers who are plagued with obstructions as a result of an overabundance of lights, signs, buildings, etc. in their geographical area. Venturing alone into Nature may result in the recognition that embracing such solitude becomes a means of relinquishing control, and opening oneself up to a primal experience

(Kahn & Hasbach, 2013, p. 211). However, the need to venture out safely (a *psychological sortie* to coin a phrase) and rush back to a "safer" environment is too much like a WWI aerodrome mindset. Some amazing scientists have even vented ways to venture out into Nature without leaving the safety of the workplace—exposing only the brain to nature and not the physical body to nature. These biofeedback technologies seem a lot like playing a flight sim instead of flying a real patrol and feeling refreshingly brave (Garten, 2011). But there is no equal to real danger, nor the same emergent bonds of lifelong camaraderie among virtual flyers as there is with genuine flyers. A true sortie out into Nature (a *biological sortie*) is best taken without a parachute and above genuine clouds.

CRIME AND PUNISHMENT MEET THERIOCIDE

The human proclivity to name a behavior he is uncertain he should do using a word which makes it less uncomfortable to talk about permeates human culture—with the consumption of beef instead of cows, pork instead of pigs, and products like pesticides instead of poison. The latter is the most curious, because most words ending in -cide mean murder (e.g. homicide, patricide, matricide, filicide, and genocide), unless you are referring to nuisance animals encountered during the normal course of business. *Naturecide* might be such a word if there were such a word for what Mankind has done to the former flora and fauna of his dying world. But *theriocide* seems to have taken hold as a way to describe the unpunishable crime of killing vast unrepresented groups of animals (Beirne, 2014). *Rat poison* and *weed killer* are much less concerned with avoiding their true nature. Even men are compared to food animals to justify their slaughter (Hodson, Kteily, & Hoffarth, 2014). Man has lost his connection to the living universe, probably because the successful conquest of Nature has required the need to dissociate himself from Nature. More than a half century ago, C.G. Jung (1964) observed that:

As scientific understanding has grown, so our world has become dehumanized. Man feels himself isolated in the cosmos, because he is no longer involved in nature, and has lost his emotional "unconscious identity" with natural phenomena... Thunder is no longer the voice of an angry god, nor is lightning his avenging missile . . . No voice now speaks to man from stones, plants, and animals, nor does he speak to them believing they can hear (p. 132).

Noe's *Theory of Biological Consciousness* raises some optimism that Jung's thunder, lightning, and the cosmos have not ceased talking to humans, but rather it is humans who have ceased listening to the messages of Nature. Although humanity may not want to hear it, Nature is communicating loudly and clearly in the language of drowning polar bears as well as the anger of increasingly more violent weather patterns—just as any drowning creature would react to being abandoned by its vessel (Bradshaw, 2013).

In 2009, Robert Greenway, a pioneer in the field of ecopsychology, suggested that the cure for wildlife extinctions and environmental collapse will not arise from humans *reconnecting* with Nature because this goal is based upon the implicit misconception that "the existence of a relationship implies separate entities or processes" (Kahn & Hasbach, 2013, p. 134) which is in conflict with the multi-cultural perspective that maintains humans including their minds, have never been disconnected from nature. According to Greenway:

Reconnections between mind and nature . . . such as gardening, diet, natural dwellings, nature study . . . transformation via vision quests and long-term wilderness immersion, though often beneficial (usually pleasurable as well) are still based upon the illusion – the initial distortion- that minds can be separate from nature. (as cited in Kahn & Hasbach, 2013, p. 134)

His observation highlights the premise that the degradation and destruction of the Earth that humans seek to heal is based upon *denying connection,* so that Western society's erroneous sense of separation has resulted in the legitimization of the existence of a psychological dissociation which was considered to be "normal." In psychology, this is sometimes called *estrangement.* Estrangement from a former spouse, parent, or children, is when humans alienate themselves from them and view them as too disagreeable or impeding of their point of view, or something that needs to be purified/reconciled to their way of doing things. Humans do the very same to one another as humans do to Nature—humans furiously "disown" their families or "excommunicate" a fellow believer who does not agree with their unflinching doctrines (Damasio, 2003).

However, at this point in time the connection has finally been made between the garbage islands the size of some states floating out in the Pacific Ocean, and the hands that toss away those plastic bags. The link has finally been forged between bulldozing the Earth and the homes of bears, deer, and squirrels to build homes for humans, so that suddenly, human actions made possible through psychological dissociation is no longer anonymous or blameless, so that humans are no longer being afforded their past mental and ethical refuge from the consequences of their actions with the realization that an ensuing global collapse is indeed personal (Bradshaw, 2013).

Still, even Greenway's breakthrough insights are a far cry from the need for a perception of Nature as part of what humans are, rather than this species-wide, tolerance-oriented *agnosia*—that denies being a part of Nature, or needs to feel what Nature feels in order to survive. For the layperson, there are a host of *agnosias* that surface in brain-injured or brain-diseased patients, which are neurological conditions in which physical objects are visually acknowledged but haptically disidentified with (any neuronal connectivity is denied)—even when they are attached to the person's own body (Smith, 2006). An example is denial of the left arm being part of one's own body, so that the denier does not care what happens to it and gets angry if told any differently. This is so very equivalent to how the brain common to the human species views every other animal in its natural world—as though something neurological has gone horribly wrong, allowing the self-concept to mistreat and even discard parts of its own biological vessel in the belief its mind does not need them. *Prosopagnosia* is one of these agnosias, and is the inability to recognize extremely familiar faces—which aptly describes the human claim that only the human face matters to its future (*anthropos*)—while angrily denying that any other face (that of a sentient animal, e.g.) matters. Humans deny that animals are conscious, alive, experience pain when horribly injured, or care about their own future or that of their families—imbuing them with a mechanical physics eerily similar to early automatons used for entertaining the rich.

With agnosia, the fact is clearly understood, but the innervation and vascularization necessary to promote sympathy along with the urge to protect the living flesh are eerily absent. Curiously, the human species *does* feel sympathy for its manufactured works and writing and defends them virulently as though living limbs on its own physical body, at risk of being amputated. A substitution of body parts for mental facts seems to have taken place mentally over the course of human evolution, like grafting in artificial needs unnecessary for biological survival and grafting out natural needs essential to biological survival . . . which phenomenon psychologists see in phantom limb syndrome as well (Meinecke, 2017; Ramachandran & Altschuler, 2009; Ramachandran, Rogers-Ramachandran, & Cobb, 1995;). Even though a living limb has been severed from the body, the body continues to listen for feedback to see why its severed limb does not respond—much the same as someone who is missing in action (MIA), or a child that disappears and is never seen again by its grieving family. Its family (the colony of cells that hosts that dismissive mind) will never stop searching for their beloved attachment, even if the mind has already moved on. This is precisely how the human mind views the severing of products from the Earth, while the Earth cannot move on as though nothing has been beheaded.

Ian McCallum (2013) adheres to the philosophy of *Wild Psychology* which purports that the deteriorating conditions, such as diminishing access to clean air and water as well as sustainable soils created by the accelerated growth of the human population, are not only detrimental biologically to all species, but also have a negative impact psychologically to all species. Considering that 50 percent of the Earth's forests, wetlands, and peat beds have been destroyed by humans over the past one-hundred years, these wild landscapes of the *soul* as well as a part of the human identity, and their innate need to belong are now in peril. Humans appear to have difficulty in confronting their evolutionary heritage of their behavior as sharing wild and non-human traits, and they seemed to have confused *wildness* with *savagery*. Historians, anthropologists, and sociologists have, for some time, admitted to a need to purge colonial thought and expansionist doctrine from the terminology used to described human groups that do not think and act as industrialized societies think and act (McCallum, 2013). Terms like *savages* and *primitives* presume that the Industrial Age is akin to an evolutionary gain that benefits the world's biologically-based ecology, whereas most evidence suggests explosive brain growth, mass production, and machinery are hastening the extinction of millions of years of successful, mutual evolution (aka mental savagery). Humans tend to prefer to think of themselves as being *civilized*; so, when faced with anything wild, spontaneous, unpredictable, and downright indifferent to human existence, humans regard it or them (referring to non-humans) as being *threatening* or *alien* which results in misunderstanding, fear, and its ultimate demise or destruction. Clearly, there is an immediate need for a radical change in human attitude toward itself and the environment if there is any chance of establishing a trans-species collaboration which is mutually beneficial for all species, rather than conditional upon the survival of just one egocentric species.

MILLIONS OF SPECIES SHARE A COMMON ANCESTOR

Thanks to the mapping of the human genome, there has been a realization that humans and non-humans share the common language of DNA where they are all related at least to a degree, and the *web of life* has become a reality. Every mammal ranging from mice to elephants shares more than 90 percent of the human genome, so that the "elephant in the room" metaphor may just refer to a long lost relative (McCallum, 2013). Every single cell in the body of every single living organism (human or non-human) contains DNA which is the self-replicating material that passes hereditary traits intergenerationally or from one generation on to subsequent generations. This information is encoded within the sequencing of only four chemical bases: adenine which is designated as *(A)*, guanine or *(G)*, cytosine or *(C)*, and thymine or *(T)*, so when discussing the shared DNA between humans and non-humans, it is this sequencing pattern that is under investigation. Scientists have suggested that while humans share DNA sequencing with *all* other living organisms, they tend to share a greater amount of DNA with non-humans who are closer to them in the evolutionary line as well as with common ancestors than with those non-humans further removed. So, in essence, the query as to whether or not humans evolved from apes, missed the point completely since humans *are* apes due to the arrangement of the biological groups and subgroups to which humans belong. Humans are most closely related to chimpanzees and bonobos with whom they share 98.7 percent of their DNA according to researchers at the Max Planck Institute in Germany. However, humans share 1.6 percent of their DNA with bonobos that they do not share with chimpanzees, and 1.6 percent of their DNA with chimpanzees that they do not share with bonobos. While the common ancestor of all mammals has been calculated to have existed more than 25 million years ago, the DNA comparisons between humans and non-humans can be quite enlightening. Research has indicated that humans share approximately 93 percent of their DNA with rhesus monkeys, 90 percent of their DNA with Abyssinian house cats, and 85 percent (on average) of their DNA with mice. Perhaps a more striking statistic relating to common DNA percentages is the fact that humans share approximately 60 percent of their DNA with a banana. Note that statistical representations of DNA may be misleading since much of the DNA is *silent* or not involved in the actual coding sequence (Deziel, 2018).

William Blake penned that humans were stardust which turned out to be a fact since the hydrogen atoms contained in the human body, trees, rivers, rain, etc. are 13.7 billion years old, and iron which is the core ingredient of hemoglobin originated from exploding stars, as do the elements of magnesium (Mg), potassium (K), and calcium (Ca). These elements combined with carbon (C) originating from the sun, are necessary for the species to successfully wage their battle for survival according to theories of *biological survival* which state that for any organism to survive, it must exist within an environment or an element in which it is able to locate or produce sufficient food to provide it with enough energy to live long enough to reproduce. The basic elements required for biological as well as psychological survival are as follows: environment (territory or element), resources (food), energy, ability to adapt, time, and sex (reproduction); however, behavioral skills or strategies are also necessary adaptive behaviors which permit organisms to establish their territories, find food, sustain energy, remain alive, and reproduced effectively. These behavioral skills are balanced by the evolutionary strategies of *fight* or *flight* which may be represented on a spectrum (moving from fight on one end to flight on the other end) to include: "challenge, competition, opportunism, risk-taking, deception, cooperation, compromise, sacrifice, withdrawal, and escape" (McCallum, 2013, p. 145).

However, there are several components missing from this equation for biological and psychological survival when spirit, morality, freewill, and play (fun, frolic, and fantasy) are not featured in the formula. Without play, the survival of particular bird species in addition to every mammal would be of a lesser quality if not downright impossible. Play is the survival compass of every mammal through which he or she learns skill-refinement activities for establishing his or her turf, the procurement of food, and sexual foreplay to attract and retain a mate for reproduction as well as the development of a testing ground for honing fight-or-flight mechanisms. A survival compass is essential for all species that are social as it is a prerequisite for constructing, reinforcing, and restructuring of alliances, relationships, and/or hierarchies. Within the human species, play is inseparable from problem-solving as well as from stretching the imagination to embrace the sublime to the ridiculous, the absurd and the outrageous, puzzles and mind games to role-playing and brainstorming in focus groups which has also been referred to as *serious play*. Perhaps the rule-of-thumb might be stated as follows: the more dependent a species happens to be upon learning by trial-and-error, through mentorship and/or apprenticeship in order to thrive, the greater that species' need for play is. A unique feature of playgrounds is that they are movable, and exist in the mind just as well as they exist in the physical world, so whether games played are won or lost, they serve as a preparation for life. Play is Nature's invitation to test an individual's physical and intellectual prowess without malice, but it functions closely as a dress rehearsal for the "kill" to acquire turf, resources, status, food, and/or a mate (McCallum, 2013).

The lesson then, in calling for a wild psychology becomes an attempt to explain and to comprehend (not condone) the underlying substrates of human behavior by viewing it through an evolutionary lens. McCallum (2013) writes of McLean's *triune concept* of the human brain, which is divided into three distinct evolutionary components (a reptilian component, and ancient mammalian component, and a modern mammalian component (p. 147). Each is associated with a kind of intelligence (reactive intelligence, responsive intelligence, and reflective intelligence) and this could be helpful as a possible model, in the rediscovery of the human relationship with the wild world. Employing an evolutionary perspective highlights the biological, social, and individual life of humans a bit differently with a reduction in the level of surprise relating to the deception, guile, and motivation or gumption underlying human behavior. The dark side of human nature which includes conflict, power struggles, territoriality, in-groups, out-groups, dehumanization, tyranny, racism, and biases will never cease, and seems to be present in each form of intelligence no matter the species, in the struggle for mental existence (Meinecke, 2018a, 2018b). However, identifying these human tendencies (sometimes called the *Dark Triad* when they cannot be concealed), and recognizing their evolutionary significance does offer the opportunity to mitigate and/or eliminate these destructive dynamics. It affords an opportunity to remove the stigma attached to certain mental and physical disorders and/or behavioral maladjustments and ask whether the human divergence from a commensal relationship with Nature toward a commensal relationship with its own thoughts underpins its disregard for Nature (Meinecke, 2017; O'Grady, 2016). Evolutionary thinking disseminates a profound ecological message regarding the healing capacity that Nature provides through her often little-acknowledged interaction with streams, trees, landscapes, animals could bring into their lives via that wild thread that intertwines all biological life. The human psyche is comprised of a mind-body, mind-Earth phenomenon as represented by a complex interplay of genes (DNA), imagination, and the Earth's biosphere and beyond. To recognize, acknowledge, and live this connection adds depth to the concept of belonging which also raises an awareness of a higher sense of synchronicity in all relationships (human and non-human) as well as in fostering a collaboration among *all* species.

ODD BONDS: BEYOND SPECIES LINES

According to Dr. Marc Bekoff (2012) in his comments that were published on *Psychology Today's* website, it has become widely accepted that non-humans (just like humans) are conscious beings who experience and display a wide variety of emotions which include "joy, happiness, pleasure, love, empathy, compassion, and sadness and profound grief" (p. 1). These experiences are really not all that surprising since they share similar structures and neurochemicals which are involved in processing and expressing their feelings. It has become more often that observations regarding strong emotional attachments and friendships across species lines have been publicized when they are shared by improbable friends including predators and prey, such as a cat and a bird, a snake, and a hamster, etc. Other examples of emotions being shared between different species are exemplified by the close and enduring relationships between humans and nonhumans with whom they live, work, and rehabilitate when necessary.

On November 7, 2012, the PBS Nature documentary, *Animal Odd Couples* aired contained some of the most amazing footage of these unanticipated and improbable relationships. The (advertising) video trailer was accompanied by the following description of the program:

Are animals capable of feeling complex emotions? Recent observations of unexpected cross-species relationships in zoos and animal sanctuaries around the world may provide some answers. Endearing interactions between a cheetah and a retriever, a lion and a coyote, a dog and a deer, a goat and a horse, and even a tortoise and a goose offer captivating glimpses of supportive connections in the animal world. Each interspecies pair challenges the conventional wisdom[1] that humans are the only species capable of feeling compassion and forming long-lasting friendships. (Nature, 2020, para. 1)

This documentary was based upon sound scientific research with the promise of future research revealing that these close interspecies relationships among odd couples are more common than was previously believed; that empathy, compassion, and kindness do indeed cross species lines; and that non-humans are also capable of demonstrating moral behavior. "Animal Odd Couples" did a great deal in raising the public awareness and appreciation that non-humans also have rich emotional lives, and experience the same emotions that humans do (Bekoff, 2012). Perhaps humans and non-humans are not living in such different worlds after all?

The authors would like to propose here a much more inclusive attitude toward the environment than this reluctant inclusion of a few human-like species in the idea of Life. A long-overdue inclusion of the rights of minority groups longitudinally excluded by the limited human rights declared to be the rights needed by all people, is not a good way to guarantee anybody's rights to equal treatment (Meinecke, 2017; Plous, 2002). Similarly, gradually including a few (human-like) species with "complex emotions" in the amazing category of sapient beings only a favored subset of which is guaranteed the right to a dignity all beings need, is perhaps the most arrogant and egocentric attitude of all. Whether planet Earth has complex emotions or not will not motivate humans to save it (because the Earth does not need to be animate to do what it does best); by dividing the simple from the complex, or the vertebrate from the invertebrate, this species has fabricated an insoluble distinction guaranteeing the right of financially motivated industries to exploit other living beings without their consent. Gradual inclusion is a means to delay inclusion of things that the dominant group has no intention of including (by delaying inclusion until the groups under consideration die of "natural causes," even though any detaining of said inclusion is unnatural). If the human species must assemble together to debate the quality of mercy, the Earth's hope of mercy is already being strained beyond the breaking point by a people that does not comprehend the idea of mercy, only ratification of opinion as a substitute.

POWER DYNAMICS

Power dynamics are present in almost every human social interaction between humans and humans or humans and non-humans according to Dr. Ana Guinote at University College, London who examines social hierarchies, "Power is everywhere" (Weir, 2017, p. 41). She and other researchers have made progress in determining how power or a lack of it, affects the way humans think, make decisions and behave. Power exists across many contexts, such as social status, influence, and wealth; and is defined by Dr. Adam Galinsky as "an asymmetric control over valued resources in a social relationship" (Weir, 2017, p. 42). Galinsky explains that when an individual possesses all the power he or she requires, he or she is not dependent upon others, and therefore, does not have power wielded over him or her; however, if that individual possesses resources that other people want, he or she is able to wield power over *them.* In a review of the literature spanning several disciplines including (non-human) animal studies, social psychology, neuroscience, and management, Guinote (2017) concluded that powerful people tended to be goal-oriented and very keenly focused on obtaining goals that they deem important to them; but they are also willing to work primarily toward their objectives since power exerts a motivating force on people. Until more recently, what was motivating these people to act or to take charge was not that evident. Where these powerful people inherently better at identifying the advantages of taking action or were they just better at ignoring the risks? The researchers, Whitson, Liljenquist, Galinsky, Magee, Gruenfield, and Cadena, determined that it is easier for the powerful people to act and take risks because they see their world as being less threatening; but in the real world, less powerful people may have a more accurate view of the situation (Whitson et al., 2013). The powerful people demonstrating a false sense of security and in being "bullet-proof," can result in the creation of situations that are not only problematic but downright dangerous and destructive for all concerned. Since the powerful people tend to maintain a laser focus on their personal goals, they may discount the needs and well-being of others, be less willing to compromise, and rely on stereotypes and mental heuristics when making decisions, Guinote also concluded in her 2017 review. According to Galinsky, Rucker, and Magee (2016), access to more power positively correlates with engaging in more egocentric behaviors while the lack of access to power positively correlates with engaging in more prosocial behaviors. Numerous research labs employing a variety of research methodologies have determined that access to power reduces an individual's ability to demonstrate empathy or to demonstrate his or her ability to view things from another's viewpoint. In addition, Galinsky, Magee, Gruenfeld, Whitson, and Liljenquist (2008) concluded that the powerful people were more likely to make decisions based upon how useful that human (or non-human) might be in aiding them in obtain their goals. A lack of, or a reduced capacity to emphasize with others, a failure to identify risks, and/or a tendency to make snap decisions is likely to result in a devastating combination for all parties (human or non-human).

Galinsky's review suggested that although power can be associated with a reduced perspective-taking or ability to emphasize with others, power might actually make it simpler to embrace other's viewpoints, if or when, the powerful people feel a greater sense of responsibility toward others (Galinsky et al., 2016). To avoid relying too heavily on the moral compasses of those powerful people, Galinsky also suggested the construction and implementation of policies and systems that will guarantee accountability. Also, in so many instances, people have an inaccurate view of their personal ability to control their lives to where they do not realize just how much power or agency they actually possess (Weir, 2017), and just much of a positive impact they are able to make in reducing the existing conflict between humans and Nature. It should be added here, that, just as the ideas of justice and mercy are far too limited when it comes to inclusion of minority human and non-human groups in the concept of conscience, so too notions such as responsibility and accountability circumvent the idea of a common reverence for Life, which would obviate the need for either reluctance or regret when applying reason and conscience to all living creatures without respect for persons or exclusion by reason of genetic make-up or infirmity.

CONCLUSION

In this chapter about losing everything, the reader learned that there is a palpable difference between deliberately losing a little, and randomly losing everything. Humans simply cannot control everything, and when humans think they can and find out they cannot, a concept called *Force Majeure* comes into full force in their lives. But that is okay, because—having no one to blame for this common misfortune, an ineffable gratitude is the sole survivor of these common losses. Mutual loss (when extreme) brings about the opportunity for unusual acts of mercy and affection toward those who had not been of concern before, yet when suddenly connected by a common misfortune, anyone's pain becomes too much pain to bear. Altruism has gotten a bad name these days, so that even its existence is in doubt. It may be that imagining an outcome to try and save oneself—rather than courageously deflecting the urge to think before the act of mercy—is more costly, less survivable, in the final analysis. Whether to be motivated by what is rational and survivable, or to scorn Reason and be guided by what is right and kind (though utterly irrational), might unexpectedly provide for an outrageously beneficial outcome even Chance had not considered.

Darwin (1876) wrote that the idea of a species is not a fact, just a convenient way to divide what is otherwise indivisibly dear and insensible. So when humans mistake conveniences for facts humans begin to think that some species are more vital than others, and sever from their futures what was part of their essential makeup today. A trans-species perspective may bypass this *endearing yet inferior* mindset so that those parts of Creation humans have been feeling ashamed of (or not in need of), will receive the greater honor not less honor (1 Cor. 10: 23-25). Are humans devolving as a species so that humans mistake aversion to taking risks to save their world for an enlightened form of rational decision making? Psychological versus biological risk-taking results in a species afraid to live Life for real - among the dangerous yet worthwhile wonders outside the mind. In addition, disconcern for other living things is a lot like a neurological disease called *agnosia*—a denial that parts of your own body belong to you, and viewing one's own nuclear family as complete strangers.

Beyond the frontier of an anthropocentric augury that confines its mercy to its city limits and its municipal jurisdiction (much as ancient Roman seers once did), there lies an undiscovered country overflowing with a wild mercy that knows no municipal boundaries. Humans should not be surprised if, like Portia in the *Merchant of Venice*, the quality of mercy holds the antidote to this civilized species' intractable malice against Nature. Maybe if humans lose everything, humans will see just how much Nature meant to them, and be thankful for what they briefly had.

REFERENCES

Beirne, P. (2014). Theriocide: Naming animal killing. *International Journal for Crime. Justice and Social Democracy*, *3*(2), 49–66. doi:10.5204/ijcjsd.v3i2.174

Bekoff, M. (2012). *Odd couples: Compassion doesn't know species lines.* Retrieved on July 18, 2019, from https://www.psychologytoday.com/us/blog/animal-emotions/201210/odd-couples-compassion-doesnt-know-species-lines

Borchers, J. G., & Bradshaw, G. A. (2008). How green is my valley—and mind: Ecotherapy and the greening of psychology. *Counseling Today*, 38–41.

Bradshaw, G. A. (2006). No longer a mind of our own. *Seed*, 43-45.

Bradshaw, G. A. (2009). *Transformation through service: trans-species psychology and its implications for ecotherapy. Ecotherapy* (L. Buzzell & C. Chalquist, Eds.). Sierra Club.

Bradshaw, G. A. (2010a). *Animal visions*. Retrieved Nov. 2, 2011 from https://animalvisions.wordpress.com/2010/09/17/trans-species-living-an-interview-with-gay-bradshaw/

Bradshaw, G. A. (2010b). We, Matata: Bicultural living amongst apes. *Spring Journal*, *83*, 161–183.

Bradshaw, G. A. (2013). Living out of our minds. In *The Rediscovery of the Wild* (pp. 119–138). Cambridge, MA: MIT Press.

Bradshaw, G. A., & Watkins, M. (2006). Trans-species Psychology: Theory and Praxis. *Spring, 75*, 69-94.

Bruneau, E., Jacoby, N., Kteily, N., & Saxe, R. (2018). Denying humanity: The Distinct Neural Correlates of Blatant Dehumanization. *Journal of Experimental Psychology. General, 147*(7), 1078–1093. doi:10.1037/xge0000417 PMID:29847980

Burns, R. (1785). *To a mouse, on turning her up in her nest with the plough.* Retrieved on March 11, 2020 from http://www.robertburns.org/works/75.shtml

Cameron, D. C., Hutcherson, C. A., Ferguson, A. M., Scheffer, J. A., Hadiandreou, E., & Inzlicht, M. (2019). Empathy is hard work: People choose to avoid empathy because of its cognitive costs. *Journal of Experimental Psychology. General, 148*(6), 962–976. doi:10.1037/xge0000595 PMID:30998038

Damasio, A. (2003). *Looking for Spinoza: Joy, sorrow, and the feeling brain.* Orlando, FL: Harcourt.

Darwin, C. R. (1876). *The origin of species by means of natural selection, or the preservation of favoured races in the struggle for life* (6th ed.). London, UK: John Murray.

Davey, G. (2011). *Applied Psychology.* West Sussex, UK: BPS Blackwell/ John Wiley & Sons Ltd.

Deziel, C. (2018, July 20). Animals that share human DNA sequences. *Sciencing.* Retrieved on August 9, 2019, from https://sciencing.com/animals-share-human-dna-sequences-8628167.html

Druck, K. (2017). *What it's like to lose everything and start over.* Retrieved on July 11, 2019, from https://www.nextavenue.org/what-its-like-to-lose-everything-and-start-over/

Galinsky, A. D., Magee, J. C., Gruenfeld, D. H., Whitson, J. A., & Liljenquist, K. A. (2008). Power reduces the press of the situation: Implications for creativity, conformity, and dissonance. *Journal of Personality and Social Psychology, 95*(6), 1450–1466. doi:10.1037/a0012633 PMID:19025295

Galinsky, A. D., Rucker, D. D., & Magee, J. C. (2016). Power and perspective-taking: A critical examination. *Journal of Experimental Social Psychology, 67*, 92–92. doi:10.1016/j.jesp.2015.12.002

Garten, A. (2011). *Ariel Garten: Know thyself, with a brain scanner* [Video file]. Retrieved from https://www.ted.com/talks/ariel_garten_know_thyself_with_a_brain_scanner.html

Gershon, L. (2019, January 23). *Chivalry was established to keep thuggish, medieval knights in check*. Retrieved on March 11 from https://www.history.com/news/chivalry-knights-middle-ages

Gorsuch, R. L., & Smith, C. S. (1983). Attributions of responsibility to God: An interaction of religious beliefs and outcomes. *Journal for the Scientific Study of Religion, 22*(4), 340. doi:10.2307/1385772

Guinote, A. (2017). How power affects people: Activating, wanting and goal seeking. *Annual Review of Psychology, 68*(1), 353–381. doi:10.1146/annurev-psych-010416-044153 PMID:27687123

Hodson, G., Kteily, N. S., & Hoffarth, M. R. (2014). Of filthy pigs and subhuman mongrels: Dehumanization, disgust, and intergroup prejudice. *TPM. Testing, Psychometrics, Methodology in Applied Psychology, 21*(3), 267–284. doi:10.4473/TPM21.3.3

Jung, C. G. (1964). Approaching the unconscious. In C. G. Jung & M. L. von Franz (Eds.), *Man and His Symbols* [Kindle Ed.]. London, UK: Aldus Books/Dell Publishing.

Kahn, P. H. Jr. (1999). *The Human Relationship with Nature: Development and Culture*. Cambridge, MA: MIT Press. doi:10.7551/mitpress/3604.001.0001

Kahn, P. H. Jr. (2011). *Technological Nature: Adaptation and the Future of Human Life*. Cambridge, MA: MIT Press. doi:10.7551/mitpress/7983.001.0001

Kahn, P. H. Jr, & Hasbach, P. H. (2013). *The Rediscovery of the Wild*. Cambridge, MA: MIT Press.

Keltner, D. (2009). *Born to be good: The science of a meaningful life*. New York, NY: W.W. Norton & Co.

Lyubomirsky, S., & Tkach, C. (2003). The consequences of dysphoric rumination. In C. Papageorgiou & A. Wells (Eds.), *Rumination: Nature, theory, and treatment of negative thinking in depression* (pp. 21–41). Chichester, UK: John Wiley & Sons. doi:10.1002/9780470713853.ch2

Manning, D. (2011). *Don't take my grief away from me*. Oklahoma City, OK: InsightBooks.

McCallum, I. (2013). A Wild Psychology. In *The Rediscovery of the Wild* (pp. 139–156). Cambridge, MA: MIT Press.

Meinecke, L. D. (2017). *Neglected by assessment: Industry versus inferiority in the competition for scarce kidneys* (Doctoral dissertation). Available from ProQuest Dissertations and Theses database. (ProQuest No. 10689852)

Meinecke, L. D. (2018a). A theory of cognitive idolatry, or the human struggle to preserve psychological certainty in a biological species - chapter 1 (Research proposal). doi:10.13140/RG.2.2.16527.43684

Meinecke, L. D. (2018b). A theory of cognitive idolatry, or the human struggle to preserve psychological certainty in a biological species - chapter 2 (Research proposal). doi:10.13140/RG.2.2.31089.53601

Meinecke, L. D. (2020a, March 7). *Nature-nurture and language development*. doi:10.31234/osf.io/4u3kq

Meinecke, L. D. (2020b, March 14). *Tripartite's fourth part: The role of faith in attribution theory*. Retrieved from psyarxiv.com/b4um5

Nakahashi, W., & Ohtsuki, H. (2018). Evolution of emotional contagion in group-living animals. *Journal of Theoretical Biology*, *440*(7), 12–20. doi:10.1016/j.jtbi.2017.12.015 PMID:29253506

Nature. (2020). *Animal odd couples*. Retrieved on March 13, 2020 from https://www.pbs.org/wnet/nature/animal-odd-couples-introduction/7958/

Norretranders, T. (2014). *2014: What scientific idea is ready for retirement?* Retrieved on July 11, 2019, from https://www.edge.org/response-detail/25500

O'Grady, M. A. (2016). *The theory of biophilia* (Reg. No. TXu 2-032-922). United States Copyright Office. Retrieved from https://www.copyright.gov/

Pets for Vets. (2020). *Building bridges*. Retrieved on March 11, 2020 from https://petsforvets.com/

Plous, S. (2002). *Understanding prejudice and discrimination*. Boston, MA: McGraw-Hill.

Prosocial Behavior. (2019). Retrieved on July 16, 2019, from https://www.goodtherapy.org/blog/psychpedia/prosocial-behavior

Ramachandran, V. S., & Altschuler, E. L. (2009). The use of visual feedback, in particular mirror visual feedback, in restoring brain function. *Brain*, *132*(7), 1693–1710. doi:10.1093/brain/awp135 PMID:19506071

Ramachandran, V. S., Rogers-Ramachandran, D., & Cobb, S. (1995). Touching the phantom limb. *Nature*, *377*(6549), 489–490. doi:10.1038/377489a0 PMID:7566144

Savage-Rumbaugh, S., Wamba, K., Wamba, P., & Wamba, N. (2007). Welfare of Apes in Captive Environments: Comments On, and By, a Specific Group of Apes. *Journal of Applied Animal Welfare Science*, *10*(10), 1. doi:10.1080/10888700701277261 PMID:17484674

Smith, A. D. (2006). Imagery. In G. Davey (Ed.), *Encyclopaedic Dictionary of Psychology* (p. 125). London, UK: Hodder Arnold.

Tulane University. (2020). *What is social exchange theory?* Retrieved on March 11, 2020 from https://socialwork.tulane.edu/blog/social-exchange-theory

Wang, S., Lilenfeld, S., & Rochat, P. (2018). Schadenfreude deconstructed and reconstructed: A tripartite motivational model. *New Ideas in Psychology*, *52*, 1–11. doi:10.1016/j.newideapsych.2018.09.002

Weir, K. (2017, April). Power play. *Monitor on Psychology*, 41–44.

Whitson, J. A., Liljenquist, K. A., Galinsky, A. D., Magee, J. C., Gruenfield, D. H., & Cadena, B. (2013). The blind leading: Power reduces awareness of constraints. *Journal of Experimental Social Psychology*, *49*(3), 579–582. doi:10.1016/j.jesp.2012.10.009

Zimbardo, P. G. (2007). The Lucifer effect: Understanding how good people turn evil [Kindle ed.]. New York, NY: Random House.

ENDNOTE

1 Wisdom: this author is in total disagreement with this word choice.

Chapter 10
First Love:
How Conflict Resolution Can Cure Indifference

ABSTRACT

The science of conflict resolution is a good candidate for a new science dedicated to resolving the age-old feud between mankind and nature, an unwarranted feud initiated and perseverated against nature without provocation or retaliation by its domestic partner. Although representative conflict resolution models exist such as ad litem, nature presents a unique problem when it comes to conflict resolution, because (unlike a human child) nature is not meant to "grow up" and become a career-focused, tax-paying citizen. Marriage counseling might work, but the genuine needs of nature cannot be reconciled with the imaginary rights of man, nor his demand to dominate and abuse his partner. A reverence-based biophilic model is needed, wherein true love matters more than the husband's prosperity.

INTRODUCTION

In this chapter, the authors will form a comparison between the science of conflict resolution and the nature of the current environmental dilemma. The relationship between the planet Earth and the human species is not unlike a monogamous affective contract gone awry. The human species has been unfaithful; the partner has had enough. But all is not lost—in any

DOI: 10.4018/978-1-7998-4408-2.ch010

relationship, putting the needs of one's partner before one's own needs is an indication of a willingness to make amends. Putting love before ambition, financial prosperity, or global power can work biological miracles previously thought to be extinct. It is time for an unfaithful species (the human species) to recommit its life to the happiness of its first love—the planet Earth. The science of conflict resolution can help point the way, but the non-verbal nature of the injured partner (Nature) makes it difficult to apply verbal solutions to cure a non-verbal problem. However, by promising to put the needs of its Beloved first (this world humans call the Earth), the human species may yet find forgiveness in the eyes of a weeping Nature, and restore happiness to a once-monogamous relationship—lost to rampant egocentrism, financial affairs, and conceptual infidelity.

OVERVIEW OF CONFLICT RESOLUTION/ CONFLICT MANAGEMENT IN GENERAL

Misunderstandings, miscommunications, omissions, change, etc. may often lead to disputes among individuals, agencies, organizations, businesses, or nations for personal, emotional, financial and/or political reasons. Conflict resolution or conflict management is often defined as a means by which the parties find a peaceful and (hopefully) mutually satisfying or mutually beneficial solution to their disagreement. Therefore, the goals of an *alternative dispute resolution* process or ADR include (Nagy, 2018):

1. Finding a solution to which all parties can agree as quickly as possible
2. Improving, rather than harming the relationship among all the individuals or members of the groups in conflict
3. Maximizing outcomes by gaining access to resources that would not be available outside of participating in the ADR process
4. Gaining an understanding of the perspective of the opposing party, with particular regard to underlying motivations which may be fueling the conflict
5. Maintaining and improving relationships among the parties to pave the way for smooth partnerships and/or alliances in the future
6. Mitigating the use of resources, such as time, energy, finances, motivation, etc. which are quickly depleted during prolonged and/or escalating disagreements (Nagy, 2018)

The time to enter alternative dispute resolution is the point at which all parties are ready to come to the table which in the field of conflict resolution is called *ripe* (for resolution). If one or more of the disputing parties are unable or unwilling to participate in ADR, there is no point for the remaining parties to begin the process without a unanimous agreement. The seven steps that must be present for the successful navigation and resolution of a conflict are as follows (Nagy, 2018):

1. Understand the dynamics of the existing conflict including the parties' interests, wants, needs, concerns, hopes, and fears
2. Communicate by employing the strategy of active listening, remain flexible and non-judgmental, empathize with the opposing party or parties to learn more about what it is like to walk in their shoes, and avoid emotional outbursts
3. Brainstorm any or all potential solutions to the conflict as to what kinds of fair and legitimate agreement(s) might be reached, and decide whether brainstorming sessions should include the opposition or segregate the disputing groups
4. Identify the best resolution to the conflict by considering win-win solutions or compromises that will provide the best results for all the parties
5. Employ a neutral third party for assistance in the ADR who is acceptable to all the parties involved in the dispute, and as experts in their respective fields, can assist in determining a standard by which to assess or measure the proposed resolution(s)
6. Explore alternative solutions to determine early on in the process as to which alternatives are the most practical, realistic, and may even be improved by additional brainstorming by the parties
7. Cope with escalating stress and high-pressure tactics which may arise when one of the disputing parties is more influential or powerful than the other, "hunkers down" into a position and refuses to budge, or even refuses to engage in ADR altogether (Nagy, 2018).

ENVIRONMENTAL DISPUTE RESOLUTION (EDR)

Oftentimes, conflict resolution processes and strategies become more finely honed to address recurring issues and challenges that consistently arise within a particular field (pun intended), such as problems relating to the environment. Professor Lawrence Susskind (2019) who is an expert in the field of environmental dispute resolution published the following 15 points that pertain to environmental negotiations and dispute resolution and employ the *Consensus Building Approach* on the Harvard Law School's website:

1. EDR can be employed during the policymaking and planning stages (upstream) or after disputes have "crystalized" due to administrative decisions, such as permitting, licensing, funding, adjudication, etc.
2. EDR is only successful if or when all the parties involved are willing to come to the table to enter negotiations
3. EDR requires a process manager, preferably a professional negotiator or facilitator who is acceptable to all the involved parties
4. The EDR parties must have the opportunity to participate in and/or approve the dispute resolution agenda
5. It is reasonable, and in many cases necessary, for the process manager to be involved in away-from-the-table activities on behalf of the group
6. EDR functions best when there are opportunities for mutual fact-finding which are managed by the process manager
7. EDR should also emphasize value-creating opportunities in contrast to zero-sum choices
8. EDR cannot substitute for statutorily mandated decision-making although it can supplement legal decisions made
9. EDR will assume different forms in different global constitutional contexts
10. EDR will rarely be precedent-setting since each dispute/conflict/decision-making process is unique to the contextual details involved
11. EDR can include opportunities for confidentiality while simultaneously meeting the requirements for transparency which can be attained by the process manager engaging in *shuttle mediation* between the parties or working in caucuses. (Shuttle mediation is when the ADR professional "shuttles" or goes back-and-forth between the parties when they are unable or unwilling to communicate directly with each other)
12. Substantial advantages can be gained by creating EDR "systems" in contrast to addressing each EDR anew
13. EDR costs do not have to be shared equally by the parties, but can be contributed according to their ability to pay without jeopardizing the nonpartisan or neutral position of the process manager
14. It is possible to assess and improve EDR efforts through in-depth case-by-case analysis before, during, and after the dispute resolution process

by independent evaluators who focus on the parties' estimation of how well the process "worked"

15. Parties who are involved in EDR should consult legal counsel, but a lawyer's presence should not impede informal problem-solving dialogue among the parties

While it appears that this is an all-inclusive list which adequately addresses the process of EDR[1], what about the interests and well-being of the surrounding stakeholders collectively known as *Nature*? How are these life forms as well as Nature herself to be represented as possibly the biggest stakeholder of all with the most to lose when they and She have already lost oh, so much. It quickly becomes evident that the aforementioned conflict resolution processes are ill-suited to addressing the interests, wants, needs, concerns, hopes, and fears of Nature and her affiliated life forms. Nature is not given to verbal contests, special interests, or subcommittees; rights and sovereignty are less important to Nature as is the right to survive as it has for millennia according to the traditions of its ancestors, without interference or advice from a verbally oriented, commercially minded, and invasively expansionist species. The human condition and its egocentric agenda are ill-suited to represent the Earth's welfare and its lack of any agenda.

FAMILY MEDIATION

Adjusting the idea of conflict resolution so that it can apply to non-human relationships when it comes to Nature, means thinking about resolving conflicts within more intimate relationships—and biophilia (love of living things) is certainly that. If so, then since the party in question is not very verbal (yet), this more closely emulates *family mediation* with the participation of a *guardian ad litem.* During the course of a family mediation, if a child is unable to speak for himself or for herself due to age, maturity, competency, etc., the issue of who is able to speak for the child often arises since both of the parents, their attorneys, and even the child's attorney may have their own agendas and perspectives. In such cases, the court will appoint a *guardian ad litem* who becomes the voice for the child, and whose sole purpose is to ensure that the child's best interests become the prime directive for the court's decision. Likewise, *anyone* who purports that they represent Nature could be said to have *his or her* own agenda be it consciously or unconsciously, due to the absence of a viable subjective model for what it might mean to conceive "welfare" as Nature might conceive its idea of welfare. Perhaps one solution might be to develop a list of questions to be considered by the parties in conflict with Nature, just as parents in family mediation are asked to consider with regard to the best interests of their child. Like the parents, if the disputing parties in the Nature conflict cannot agree on how to satisfy the child's best interests, then an expert in the relevant biophilic issue could be assigned—which would be similar to assigning a custody evaluator in the child's case.

Note that these authors are not suggesting that Nature is a child or ward of the human species. Nor do the authors by any means suggest that non-human species are "less human" (childlike, childish, immature, or incompetent), simply because they are not human (Hodson, MacInnis, & Costello, 2014). However, there is a particular challenge involved in determining how to verbally represent the non-verbal splendor of Nature as a collective verbal need, let alone devise a way of representing her in a conflict or a dispute wherein verbal arguments are required to evaluate the merits of Her appeal.

MARRIAGE COUNSELING

It may even be more appropriate to address the issue of representation of Nature, not as a child's rights needing to be heard (*guardian ad litem*), but a spouse's rights needing to be heard (a suffragette's appeal against *coverture*). Coverture was a legal concept in which a woman's rights were essentially ceded to her husband's authority and protection, which is very much the way "land" was (and is) treated by "a landowner" (even if they both enjoy the fruit of that union). A mutual contract between two equal interests, however, should not require that either party cede its sovereignty to the other to enjoy a common benefit and a mutual protection. When this happens, the needs of the major partner supersede those of the minor—and this is also the case with Nature. But in an affective contract, the vows made by each party suggest a commitment to putting the needs of *the beloved other* before one's own needs—not one's own interests ahead of your beloved's well-being. Notwithstanding, this is exactly the kind of contract people make with Nature. The rights of "the People" not only come first, those rights exclude any right of the injured party (the Earth) to enjoy adequate compensation when the people abuse the rights off those they exclude from their definition.

Many sources of psychological advice seeking to aid ailing marriages, suggest putting the needs of the injured party before the needs of the party that has done the injury (Mount Vernon Therapy, 2019). The simple vouchsafing of priority (for Nature's needs) as a first step in reconciliation with Nature, would be a symbol of affection and intention to honor that vow. Following that step, environmental conflict counselors could help guide the Man/Nature partnership toward a (measurably) satisfactory therapeutic ending, one in which the concepts of reconciliation are applied and the laws themselves rewritten to remove this environmental *coverture* from future litigation perspectives—much as women's rights have been established with great difficulty during similarly adamantine eras opposed to the inclusion of minority groups in the concept of justice enjoyed only by the few.

NATURE'S RIGHTS

The *Global Alliance for the Rights of Nature* or GARN (2019) is a network of individuals and organizations that have committed themselves to the universal adoption and implementation of legal systems that are dedicated to recognizing, respecting, and enforcing the "Rights of Nature." GARN is devoted to honoring Nature and recognizing that the Earth's ecosystems—comprised of forests (trees), oceans, mountains, and animals—have the same rights as human beings. GARN is dedicated to balancing what is good for non-human species and good for the planet overall with what is good for humans. It assumes the biophilic perspective in recognizing that *all* ecosystems on the Earth are inextricably intertwined. In contrast to treating Nature as property under the law, GARN acknowledges that Nature in *all its forms,* has the *right to exist, persist, maintain, and regenerate its vital cycles* (p. 1), and further suggests that humans (people) have the legal authority and responsibility to enforce these rights on behalf of the ecosystems which can be named as the injured party or parties, with their own legal standing and rights in cases (lawsuits) that allege rights violations. In contrast, indigenous cultures who have long recognized the rights of Nature without the need for lofty adjudication proceedings, have lived in harmony with Her according to their native traditions, in which decisions and values have been based upon what is equally good for both the inhabitant and the habitat. For millennia, legal systems around the planet have treated Nature and its assets ("land" and its "natural resources") in general only as "property," where contracts, regulations, and laws have been created to protect the property rights of (human) individuals, corporations, and other entities that lay claim to living systems and their modest appeals which continue to be denied equal representation in human courts. As such, these (so-called) environmental protection laws actually legalize environmental harm by permitting (or mandating) the levels of environmental destruction and pollution that are allowable under the law without regard for Nature or her non-human species. So long as the proper studies are conducted, and only with a view to losses of natural resources relative to human interests in those sovereign resources, the impact of human encroachment on Nature's indefeasible rights to its own property and to be sovereign of its own destiny remains unaddressed. The hope of filing an appellate brief on behalf of one-sided judgments, wrongful use, and corporate malfeasance is not within Nature's foreseeable future.

There are a growing number of communities in the United States and in Ecuador who are now basing their environmental protection systems on the premise that Nature has *inalienable rights* just as humans do. This is a radical departure from the belief that Nature is only a form of human property under the law (much as certain colors and genders of humans have been viewed as "property" under the law in past eras, i.e. to legitimize slavery or provide miners of key minerals with recreational sex). However, GARN's (2019) perspective does not go far enough to grant equality to Nature under the law. It does not go far enough to encompass the breadth of the concept of protection of the non-human environment for its own sake. Instead it perpetuates an assumption which suggests that Nature needs human protection when in the final analysis, every evidence suggests Nature needs protection *from* human protection. Left to her own devices, free of human interference, She is quite capable of healing and caring for herself; yet the perspective of the human species is that Nature is a helpless child in need of parental nurture and supervision (rather than the reverse which seems amply in evidence). It seems more feasible to return Nature's rights back to Nature's peoples, rather than trying to mitigate the obvious infeasibility of applying human-oriented rights systems to non-human systems.

RIGHTS *TO* NATURE VS. RIGHTS *OF* NATURE

In 2014, an attendee at the World Parks Conference by the name of Gator Halpern (no, the author is not making this up), organized and presented at the session of Green Justice. It was designed as a forum to discuss environmental justice issues as they related to theoretical policy interventions as well as local practical applications to bring about a more environmentally just world. Well-known environmental attorneys and academics presented on how certain countries have constructed laws explicitly stating that their citizens have a right to live in a clean and healthy environment. Halpern himself focused on the conversion of *brownfields* (dumps, abandoned factories, empty warehouses, vacant lots, etc.) into public parks. Typically these types of run-down sites are located in low income communities where access to Nature and parks is limited if it exists at all. Brownfields are categorized as being two environmental injustices: first, the lack of access to parks and Nature for lower income individuals, and second, the unequal burden of living adjacent to brownfields which can be reversed by converting brownfields into green fields or public parks.

During the forum, the discussion of human's rights to Nature and access to a clean environment, shifted to a discussion of the rights that Nature Herself has (not having to live next to barren human developments), but which were not previously considered by Halpern (2014) himself. Humans may clearly state what they are allowed to do, and define the lines that even governments should not cross; however, what about extending such rights to Nature? Do the birds whose migratory patterns and indigenous airspaces were implicitly theirs long before Mankind, have a right to fly without undue risk over their own native habitats and along their traditional migratory paths? Do bodies of water have the right to aggregate unhindered without being dammed up and suckled by token lawns and hungry power facilities? There are some humans who believe that they should. For example, New Zealand in particular, has employed the rights of Nature to preserve their important ecosystems, and the Kiwi government has gone as far as to assign legal personhood to the Whanganui River by granting it "rights and interests." Perhaps the legal system is just catching up with what the indigenous Maori people have known all along by thinking of the Whanganui in that way as evidenced by them stating: "Ko au te awa, Ko te awa ko au ~ *I am the river and the river is me.* The legal agreement awarded to the river recognizes that the Whanganui and all of its tributaries are a single living entity with ownership rights over its riverbed. Since Walmart and Halliburton were given rights as human beings, is it so unreasonable to suggest that Nature be granted the same? (Within the legal system, the business structures of Walmart and Halliburton were granted the same rights as human beings have in court). Perhaps what is most disconcerting about Halpern's comments, is the fact he, as one of the members of the environmental field, had not considered the biophilic perspective of Nature's inalienable right not only to exist but to flourish.

"GREEN SENSE"

Humans' inextricable connection to Nature extends far beyond legal justice and responsibility according to Carli (2018) who suggests that everyone possesses a deep sense of ecology. This *green sense* may be defined as an intense physical, emotional, and spiritual bond with the diverse ecosystems and forms of life with whom humans share the Earth's biosphere. It is a physiological sensory perception (PSP) that helps to generate those "warm and fuzzy feelings" that are promoted by touching hardwood floors, growing plants in office environments, sharing homes with cats and dogs, and/or camping out under the stars. It is those feelings of distress that erupt when a return to a previous family dwelling has been transformed into yet another unnecessary or unwanted shopping mall. Wild and healthy ecosystems do more than nourish humans psychologically, they also provide food and shelter for a wide diversity of non-human species as well as to ensure that the Earth's air and water is purified and clean. So, for reasons such as these, each time the rights, and ultimately the health of Nature are violated, the rights and health of all species experience the same fate. Ecosystems are not fungible; the destruction of one ecosystem so it can be replaced by another ecosystem is not effective. This is evidenced by the practical implementation of these tactics around the world (Mumta, 2014). Even Steinbeck writes about the transformation of the Salinas Valley to support subsistence farming (later industrial farming), and the resulting extinction of most of the flora and fauna of the Central Valley of California (Steinbeck, 2003).

The concept of recognizing the rights of ecosystems as well as planetary life-support systems rather than regarding them as natural resources for human exploitation actually predates European colonization, enslavement, and mass genocides of the indigenous tribes of the Americas. The *Pachamama* perspective holds that human beings are only one humble member of Nature. This Pachamama worldview is enshrined in Ecuador's constitution which recognizes Mother Earth as an entity with a right to life. When Ecuador rewrote its constitution in 2007-2008, a chapter was included that was dedicated to the rights of Nature with "we the people" as being the legal authority with the power to enforce these rights on behalf of the ecosystems as well as with the ecosystem being able to be named as the defendant in any lawsuit. In a similar fashion, the Bolivian constitution guarantees that everyone has a right to a "healthy, protected, and balanced environment" which was given to both individuals, collectives of present and future generations, and to other living (non-human) things to allow them to develop in a normal and permanent fashion. Evo Morales, the indigenous-born President of Bolivia, championed the construction of the *Universal Declaration of the Rights of Mother Earth* in 2010 following the disappointing failure of the countries that attended the COP15 climate change conference to come to a consensus. This Universal Declaration was featured at the United Nations Harmony with Nature Dialogue the following year with the recognition that the usual anthropocentric, exploitation-based economy was detrimental to the life and liberty of all beings—human or non-human—on Earth. The affirmation of this document to ensure peace, justice, and a decent quality of life for all species at the Rio + 20 Conference on Sustainable Development in 2012 continues to be expanded and codified into an international law via ongoing agreements and resolutions (Carli, 2018).

It is the fervent hope of these authors, that as this process of building a global civilization founded upon the recognition that biodiversity and environmental health underpin the very health and survival of the Earth itself, there will be the realization that benefits extend far beyond the economic and into the very physiological, emotional and spiritual fiber of the web of life. By all means, nurture that sense of "deep ecology," by not resisting that urge to smell those flowers, tickle that plant, hug that tree, or succumb to whatever urge arises to fondle and adore Nature.

THE BIOPHILIC MODEL OF CONFLICT RESOLUTION

According to Mumta (2014), the laws, such as the EC Habitats and Species Directive as well as the Wildlife and Countryside Act in the UK, and the Endangered Species Act in the U.S. are designed to protect the species that are already deemed to be at-risk. However, in this inextricably connected web-of-life where scientists are reporting the extinction of dozens of species daily (which translates to a species loss of approximately 1,000-10,000 times the background rate), this protective approach is no longer sufficient. The procedure for updating the lists of endangered species takes years due to the length of time required for scientists to support their claim that a particular species is under threat. By that time, it is already too late. (Nature cannot wait for environmental impact studies to be funded and conducted before something is done to protect Her). In many countries, their (so-called) *modern law* is predicated on outdated paradigms that include the following:

- *Mechanistic* – a world comprised of separate and unconnected objects that function in a predictable manner
- *Anthropocentric* – a world existing only for the use of humans where "natural resources" and "natural capital" abound, and where "value" is based upon its utility to humans in contrast to its *intrinsic* value
- *Adversarial* – a competitive and/or retributive model where one of the disputing parties "wins" at the expense of the other party (humans *winning* at Nature's expense)

The irony of the adversarial paradigm is that in harming and destroying Nature, humans are ultimately harming and setting themselves up for destruction. The laws and cultural attitudes overall fall short of acting on this premise even when purporting a dedication to the prospect of "sustainability" for future generations. The idea of corporate sustainability (the financial welfare of non-living concepts) is not aligned with the idea of biological liberty (the welfare of living things, wherein no financial interests exist to sustain). While these *rights of Nature laws* envision a world where the laws will facilitate a global culture that respects the inter-existence and inter-dependence of all species that share the Earth's biosphere, they simultaneously create a legislative framework in which Nature is viewed as an equal stakeholder and an active participant in the maintenance of the health and well-being of the planet. This type of law differs from a *Law of Ecocide* and the *Eradicating Ecocide campaign,* etc. because it alters the "Nature as property" viewpoint. An Ecocide law would categorize extensive destruction of the environment as a crime against peace, where the focus would be on limiting harm to the environment by the use of criminal law. Any level of destruction falling short of the legal definition of what constitutes "ecocide" would revert back to the framework of environmental law, which classifies Nature as "property" thereby creating a back-to-the-drawing-board scenario (Mumta, 2014). The scope of the Rights of Nature laws and related discussions are being constructed through judicial decisions that will uphold them as well as by issuing legal arguments for the Rights of Nature. Resorting to the arguments that support the idea of Nature being granted legal *personhood,* may result in the practical applications being degraded into a litany of catch-phrases wielded by politicians, activists, etc., such as *sustainable development* and *green economy.* These may result in the justification for all sorts of decisions that have economic motivations rather than mutual benefit of all species at heart (Berros, 2017).

Perhaps a more appropriate designation would be an indefeasible attribution equally applicable to any species known as *Naturehood*, which natural right would not allow the designee to be inured to its eventual exploitation.

As attorneys, environmentalists, academics, etc. have collaborated on a European Citizen' Initiative (ECI) to petition the European Commission to construct a legislative proposal to the European Parliament and the Council of the European Union to assume a rights-based approach to environmental protection, these authors ask whether this rights-based approach just begins to address this proposed biophilic model of conflict resolution. In 2014, at the Opening of the Rights of Nature Tribunal, Casey Camp Horinek who is an elder and councilwoman of the Ponca Nation of Oklahoma, so eloquently delivered the following message:

To restore Mother Earth – her Nature's balance, the world needs to shift from a philosophy of control and dominion over Nature, and its legal system of property rights regimes, to a relationship of understanding and respect for the Natural Laws and love for the beauty of the creative female energy of Mother Earth.

To these indigenous people, Mother Nature's laws are inherent, so that any (human-made) law that denies these fundamental rights to any species is illegitimate, and therefore, a violation of all of the Natural Laws of Creation (Horinek, 2014). Presently, humans have the unique opportunity to dispense with outdated and ill-advised legal and political paradigms that have never functioned effectively, and to adopt a biophilic paradigm that is based upon a more indigenous thought and philosophy which honors an inextricable relationship among all life, all species. And so, by embracing a biophilic paradigm, the need for the rights-based paradigm is mitigated and/or eliminated completely—which also serves to reduce the (seemingly) never-ending political and legal influences on decisions pertaining to conflicts or disputes that involve Nature. The very idea of documenting the rights of everything that exists, let alone analyzing a discrete infinity of conflicts, defies the effort to conceive of any politics or jurisprudence whose careful clauses might reach the halls of litigation.

As much as these authors would love to suggest that adopting the biophilic conflict resolution model will solve any or all disputes involving Nature, it is indeed a work-in-progress with innate challenges that arise from the rules, regulations, logistics, and dynamics of previously employed conflict resolution models. For example, conflict resolution practitioners as experts in their field, are required to remain *neutral* and *unbiased* as they perform their role in an *alternative dispute resolution* process or ADR which can include negotiation, mediation, facilitation, etc. However, if an ADR practitioner assumes his or her role in an ADR process, how can he or she imply that there is no personal agenda present in his or her involvement? (Even the desire to aid Nature is, for all intents, a bias). Also, how can Nature truly be represented "at the table" during conflict resolution when most ADR processes are convened in a building, such as a courtroom, conference hall or other man-made structure? Perhaps this specific problem might be addressed by taking the issue out *into* Nature to the location which is under dispute to allow her to "testify" on her own behalf? This might just be the point at which only the phrase, *Awed by Nature* will suffice—because no words are sufficient or eloquent enough to represent the lasting majesty and fleeting glory of Nature. In Eliot's clever terminology, Nature's secret name remains *inscrutable* (Eliot & Scheffler, 2009). Perhaps this is truly the point at which the biophilic conflict resolution model has been enacted when there is no longer a need for conflict resolution as it pertains to Nature.

It may also be wise to note that, *Mother Nature does not negotiate. . .*

CONCLUSION

There is a psychological science known as conflict resolution, and this might form a model for any form of guided reconciliation between parties that have difficulty getting along. When applied to environmental dispute resolution, however, even though the principles seem systematic and sound, the interests of the key stakeholder (Nature) cannot be adequately represented by a model longitudinally evolved to represent only human entities and their interests. Further complicating matters is the problem of requiring innate or learned skills such as literacy and language to merit representation—when the injured party has always been and always will be, by nature, a non-lingual system, more in need of a common mercy than a common language.

Family mediation and the concept of *guardian ad litem* might allow the current model of child representation to play substitute to a stakeholder not too different than a protected child (Nature). In such cases, someone is appointed to act on behalf of the unemancipated minor. Unfortunately, this implies Nature is childish, immature, or incompetent, rather than that no legal models exist to represent the right to Life in all its kinds. Marriage counseling might work, since the outdated concept of *coverture* seems to symbolize the subsumed rights of a spouse who must forfeit her right to determine her own future and that of her immense holdings, in order to receive protection under the law. Yet this would only work, it seems, if (inevitably) the same laws guaranteeing her right to circumvent the law could be repealed when the time came to exceed (and therefore) repeal them.

More and more regions are trying to treat the interests of Nature as the right to enjoy the same privileges as any human of sufficient age. Still, the solution does not seem to be the addition of more rights for more beings until all beings are included under the law, but to exclude no being under the law. A biophilic model of conflict resolution waits at the frontiers of environmental justice for the chance to appeal its cause to a humble people, one which is neither partial nor impartial, but whose indefatigable reverence for Life outweighs the gavels of justice and obviates the need for an appeal.

REFERENCES

Berros, M. V. (2017). Defending rivers: Vilcabamba in the south of Ecuador. In *Can nature have rights? Legal and Political Insights*. Retrieved from http://www.environmentandsociety.org/sites/default/files/2017_i6_final_hw.pdf

Carli, J. (2018). *Legal rights for nature: A revolution whose time has come*. Retrieved from https://medium.com/@jamescarli/legal-rights-for-nature-a-revolution-whose-time-has-come-fd11c65545cf

Eliot, T. S., & Scheffler, A. (2009). *Old Possum's book of practical cats*. Boston: Houghton Mifflin Harcourt.

GARN. (2019). *What is rights of nature?* Retrieved from https://therightsofnature.org/what-is-rights-of-nature/

Halpern, G. (2014). *Rights to nature vs. rights of nature*. Yale School of Forestry & Environmental Studies, F&ES Blog. Retrieved from https://environment.yale.edu/blog/2014/12/rights-to-nature-vs-rights-of-nature/

Hodson, G., MacInnis, C. C., & Costello, K. (2014). (Over)valuing "humanness" as an aggravator of intergroup prejudices and discrimination. In P. G. Bain, J. Vaes, & J. P. Leyens (Eds.), *Humanness and Dehumanization* (pp. 86–110). New York, NY: Psychology Press.

Horinek, C. C. (2014). *Opening of the Rights of Nature Tribunal. Rights of Nature & Mother Earth. Rights-Based Law for Systemic Change*. Retrieved from https://www.ienearth.org/wp-content/uploads/2017/11/RONME-RightsBasedLaw-final- 1.pdf

Mount Vernon Therapy. (2019). *Seven signs your partner makes your relationship a priority*. Retrieved from https://www.mountvernontherapy.com/putting-your-partner-first/

Mumta, I. (2014). Being nature – Extending civil rights to the natural world. *Ecologist, The Journal for the Post-Industrial Age*. Retrieved from https://theecologist.org/2014/apr/24/being-nature-extending-civil-rights-natural-world

Nagy, J. (2018). *Section six: Training for conflict resolution*. Retrieved from https://ctb.ku.edu/en/table-of-contents/implement/provide-information-enhance-kills/conflict-resolution/main

Steinbeck, J. (2003). *East of Eden*. New York: Viking.

Susskind, L. (2019). 15 things you need to know about environmental resolution. *Program on Negotiation, Harvard Law School, Daily Blog*. Retrieved from https://www.pon.harvard.edu/daily/dispute-resolution/environmental-negotiations-and-15-things-about-environmental-dispute-resolution/

ENDNOTE

1 e.g. the role of the process manager (negotiator/mediator/facilitator), the responsibilities of the disputing parties and stakeholders, and the public or collective interests.

Chapter 11
How Reverence for a Flower Can Save a Planet

ABSTRACT

It remains a major problem that only certain things seem to matter enough to be worth saving. Many are not even on the list of endangered species, much like humans missing from the organ transplant waiting lists. If humans do not see their commercial value (profitable species), humans do not seem to appreciate them much at all (nuisance species). Psychologically, the human species tends to forget how much it loves what matters, and accidentally lets them perish (living things). Meanwhile, it falls in love with things that don't matter and can't die (writing, ideas, and cities). But if this species could only learn to have a deep and abiding reverence for a little thing (like a flower), perhaps it could learn to have as much or more reverence for its human neighbors. This chapter will go into intimate detail on how this may be practically applied. This practical application of reverence for life that stems from work with children and animals is shown to be effective in reducing intergroup prejudice. Reverence for all life can be a core and crucial learning concept for early childhood education.

DOI: 10.4018/978-1-7998-4408-2.ch011

INTRODUCTION

What though the radiance which was once so bright
Be now for ever taken from my sight
Though nothing can bring back the hour
Of splendour in the grass, of glory in the flower . . .

William Wordsworth

How can reverence for a flower save a dying planet, if the best minds from the best civilizations cannot? Why should anybody care about flowers when there are so many humans at risk of extinction? Well, it's about caring enough about the littlest things to realize that the big ones (like the planet) are worth saving too. It remains a major problem that only certain things seem to matter enough to save them from extinction, and it isn't easy to get on that list of things to save. When people disregard the least of things to save the greatest, those people often find out later on that the things they disregarded at the time were just what were needed to save them later (Psalms 188:22; Matt. 21:42; Mark 12:10; Luke 20:17; Acts 4:11; 1 Peter 2:7; Wells, 1898). Those unpredictable variations that Nature nurtures do not seem to be mistakes after all (Meinecke, 2017, 2018a, 2018b, 2020).

This author's dissertation was about a similar list of things that hoped to be saved (human lives, when replacement organs are scarce). He learned that not many people qualify for the list, and among the few that do qualify, only a few will live long enough to compete for the few organs available when their turn finally comes around (Meinecke, 2017). That waiting list for an organ is not unlike the endangered species list; most creatures never make the list, and most die long before it's their turn to be protected. In fact, there is a lot more competition to get rid of the list than to put more species on it (Knickmeyer, 2019). As jobs become more precious than life itself, economic growth seems more precious than biological growth itself.

This chapter is about rethinking what is precious enough to save—precious enough to put at the top of the list of endangered things. Whatever is precious comes first and matters most; whatever is precious can usually tell (without words) that it is honored above all else. Life is the most precious thing humans know of, isn't it? So, if humans honor Life above everything else, maybe this will help humans save their planet? Okay, but why is Life so precious? What makes a thing precious? Well, it seems true that whatever is most rare in all the world remains precious no matter where it is, and no matter where one is with respect to it. It also seems true that what cannot stay (when everyone wishes it would stay a little longer) remains precious whether it's here today or whether it left yesterday (Marchant, 1646). Whatever is precious is first, and rare—and cannot stay.

Yet humans—the dominant species on the planet Earth—are anything but rare, and humans refuse to go away. Humans come first (Leiss, 1972; Meinecke, 2018a, 2018b). Humans think they are more precious than the planet they live on (even if it does not seem logical that they can survive without it). Humans think they are more precious than Life itself (again, even if it does not seem logical to survive as a physical entity without a physical entity). Perhaps the reason the best minds cannot save the planet, therefore, is because humans have accidentally put the survival of their own species above the survival of the planet they live on? Humans may be the greatest *species*, but perhaps favoring the greatest *species* is not the way to save a favored *world.* To exclusively honor and preserve just one species, when it is teeming with a discrete infinity[1] of species, suggests a very narrow idea of honor. Perhaps a simple change of reverence (for the least of species, not the greatest of species) is in order? (Matt. 6:24-34).

REVERENCE: WHAT IS PRECIOUS?

The first thing needed in this chapter, is to try and understand how something becomes precious. This can be done that by asking a few questions. Is the most precious thing the thing at the bottom of a list of things to honor and to save? No. One puts the most precious things at the top of the list. And when one has lots of something, does one worry about losing a little of it? No, usually if one has lots of it, who cares if one loses a little? Lastly, does one worry about losing things that will always be there? No, one worries about losing things that might only be here for a little while.

There is a phenomenon called the *overview effect* which is an extraordinary feeling of wonder that astronauts report feeling when they see the planet Earth for the first time in space (De Luce, 2019; Shaw, 2017). Nobody seems to expect to spend quite so much time looking out the window at an object they know so well. But each astronaut reports experiencing a sort of a *paradigm shift* in their attitudes toward what should survive on this familiar blue ball. Suddenly the survival of national boundaries is not an issue, for none exist; suddenly these petty squabbles over territory and resources pale in significance. Suddenly humans realize how tiny and fragile their planet actually is—and it took an experience of *Dasein* (being there) to resurrect that habituated feeling that their conquest of and knowledge of the planet took away (Haugeland, 2013).

The capacity to view your own body from a distance is a very eerie experience, and the overview effect of all life on the planet from a distance is not so different (Lenggenhager, Tadi, Metzinger, & Blanke, 2007). Often the sense of timelessness and wonder is accompanied by a profound sadness, as the belligerent level of human detail coalesces placidly into a single sapphire interest—and the issue of sovereignty becomes an inessential concern relative to the fragility of Life itself (De Luce, 2019; Vrobel, 2011). What matters most of all, what is more precious than anything else, is suddenly everything out the window of your spacecraft . . . enshrouded by a semi-permeable blue and white veil that eerily separates everything living beneath the veil from a lifeless darkness just beyond that veil. Words apparently cannot begin to compete with that ineffable sensation, as though a vestigial awe for Life had been resurrected while orbiting one's own non-living biases.

REVERENCE IS ABOUT WHAT MATTERS MOST

First of all, what's precious is what comes first and matters most. This may be anything depending on what you hold in the highest regard, or what you are most afraid of losing. It's that one thing you cannot live without; you would sell all you have to have it, give up everything to get it back. Firstly, then, what is precious is what matters most to you.

What matters to most people? Well, this is not just the first question to ask, this is the first problem to consider too. If people do not see its relative commercial value, or its imagined scientific value, or its obligatory contribution to the *means of production*, that thing is unlikely to matter at all (let alone matter the most). Human civilization sees its world as a free and natural resource; human civilization sees human beings as a free and indentured labor force; human civilization sees human products and social production as everything else's purpose for existing (Deleuze, Guattari, & Foucault, 2009). This does not sound very reverent. Do human products matter most? Does human labor matter most? What about the planet? What about a view of the human species as only one among an endless diversity of precious animal species, none of which are indebted to Society from birth like human beings are?

It is as though humans look at their living planet (and themselves) as needing to possess certain ideal attributes before they appreciate them. If it does not possess some familiar visage (Shelley, 1817) or a set of manifest yet clandestine attributes people think make humankind special and every other kind less special (Leiss, 1972), people do not seem to appreciate a living thing much at all. Even if humans grant their regard to a thing in need of that regard, psychologically humans tend to *habituate* what they love physically, and end up protecting a concept instead (because a concept isn't at risk of physical death like an individual, or extinction like a species; Meinecke, 2017). Humans think what matters most is invisible and imperishable (like a human thought)—even if a thought exists only in their heads (Frosch, 1970; Rogers, 1961).

But let this question go completely the other way, all the way to the other pole. Rather than hold that the greatest things matter most (since anyone can see how great they are), what if some were to regard the least of things with as much awe as that? The problem with adoring the greatest is that there is so much competition to be the best. Whatever happens to be the best at something varies from day-to-day (or refuses to vary at all). So, it's hard to remember which thing to revere most, and everything else becomes inferior in the meantime. If people could only learn to have a deep and abiding reverence for a little thing, wouldn't a great thing realize how much more people adore it? If only people could kneel down to and lavish praise upon something as simple as a rose—perhaps those same people could learn to have as much or more reverence for one other—until they can come together again? (Luke 12:27-28; Meinecke, 2017).

A rose in the natural sense is a brief thing, a wild thing, a thing that has never been plucked or transplanted, dethorned or arranged. A rose in not a nameless stem among a bunch of roses either—as though you must have a dozen or more roses before the giving of a rose to your beloved elicits affection. Each individual flower hopes to be perceived as a stunning, *Darcey Bussell* rose; each one wants its velvet petals to be fondled (Fine Gardening, 2020). Famous thinkers like Antoine Saint-Exupéry (1943) asked that folks pause and reflect on life and consider whether the purpose of living at all is to go back and find and cherish that one rose you've had all along. Yes, there are other roses; but all they do is make you miss the one you fell in love with—and realize that she is the only one of her kind. A planet like no other is a lot like a flower like no other, when you think about it. This blue and white rose is what matters most in all the universe (Saint-Exupéry, 1943).

Sadly, the need to produce and produce to secure the welfare of just one species—when there are so many species in need of welfare—has taken a toll on *The Heart of Man* (Fromm, 2010), and robbed a once feral Nature of its indomitable spirit and a once imprescriptible freedom (Fromm, 2013). Another trio of French philosophers had a less than glowing phrase for this struggle over the means of production—they referred to it as *the law of the production of production* (Deleuze et al., 2009). There is never enough to go around, so everyone is busy producing various things to make enough things to go around (even if most of the things humans produce are things this infinitely ancient universe did not see fit to create, let alone mass produce in case it ran out of stock of those commodious items).

Little things matter—that is what this section is about. They matter because it is essential to appreciate the least of things in order to appreciate each of them in their infinite combination; and when a greater thing sees how much you appreciate a little thing of which it is comprised, it cannot help but see how much more you regard it (Matt. 6:24-34). But if one notices only the most magnificent things (the princes of industry) it is harder for the modest things (those who only stand and wait) to see where they might ever matter (Milton, 1650). A woman who was a pearl of great price to the man she married (but who never *told her* how much she meant to him), might wonder at the end of her life whether she had ever mattered to the world, or lived her life in vain (Frankl, 1990; Meinecke, 2018c).

REVERENCE IS ABOUT WHAT IS UNLIKE ANYTHING ELSE

Second of all, what is precious is also rare; what is precious is one-of-a-kind. Think about it. When you have a lot of something, do you mind losing a little? But when you have just one of something, can you lose it and still have something left? No—you would leave all you have to go and find it (Matt. 18:12). Humans have this odd idea that nothing unique exists. For everything they see, they think there is another they cannot see, and leave the one they have to find the one they don't have. Humans think all they have to do is study deeply enough to discover that this amazing new thing they just met is not unique after all. There is always something humans already know which explains what they don't know. Yet 2,000 years ago, a teacher said a good shepherd would leave a group of sheep to go and find just one. That is love; that is the kind of reverence needed to save the world—the willingness to treat each and every member of a species as a species. That worldview suggests some folks see each and every one of their charges (or constituents) as utterly special—unlike anything else under their charge. Such shepherds do not view their flocks as nameless employees producing named products, nor calculate whether some magical quantity of their people agree with this magical policy or that.

A loving husband sees his wife as the only woman in the world, as every woman in the world, as the light of his world (Meinecke, 2018c). A woman is a man's country, and he would do anything to save her. Experiments have shown that, when a male is attached to a certain female, he will actually be *rude* to potential female mates even when his primary attachment is out of view (Scheele et al., 2013). This is measurable using a chemical known as *oxytocin*. An attachment is someone revered above all else; when the two are apart, though they possess unique purposes, they cannot wait to get back together and share their *united purpose*. When a thing is rare, no matter what that thing happens to be, somebody will look at it as very precious, drop everything else, and go after it (Matt. 13:45-46). Now the finder and the found are of one purpose.

This idea of rarity and how associated it is with the unique worth of an individual has been explained thoroughly in chapter 3 (*How Taxonomy Steals Our Reverence*). The gist of that in summary, to refresh the reader's memory, is simply that it is very unlikely that Nature does anything twice which doesn't work the first time or makes two of something essential so that it finally has something essential. Remember the *Sorites paradox*? If it takes more than one thing to be a thing, that is not a very efficient way to run the universe. It is far more likely that, because humans are limited to the same abilities as a species, they cannot distinguish as individuals what they are limited to distinguishing as a kind. So, in order to pass down knowledge, humans find it more convenient to pretend there are a limited number of unique things with a limited number of behaviors, rather than attempt to learn and pass down an unguessable number of unique things with an unguessable variety of behaviors (Darwin, 1876; Meinecke, 2017). This works very well for scientific inquiry, however—enough to get by. The paradox of getting by with this convenient heaping heuristic is that it steals humanity's awe (and it doesn't have to steal that awe). It is very hard to experience an overwhelming reverence for something when one thinks it is very common. So the price of cognition, frequently, is the loss of reverence for not-so-handy stimuli in exchange for this handy reverence for cognition (Meinecke, 2017).

But if every environmental stimulus were a *wow-what-the-heck-was-that* stimulation, would humans in response to that awe treat the world the way humans do—as a huge unconscious pantry full of unstimulating commodities? Since there may only be one of each thing, and no rewind button to see what would happen if people had not exploited things the way people do, people can best express their reverence by assuming each thing is the only one of its kind. If women like men to treat them that way, why wouldn't a rose appreciate it, or a mountain, or a sea? If matter is made of the same stuff, why wouldn't it all experience a mutual wonder for its existence, whether an animal by nature, a plant by nature, a mineral by nature, or an element by nature? (Pais, 1979; Sagan, 2019).

REVERENCE IS ABOUT WHAT CANNOT STAY

The third thing about precious things is how brief they are. People don't often miss a thing they frequently see. But when there is just one chance to see it, and people know it won't last for long, it becomes the most precious thing at the time—no matter what they happen to be doing or whom they happen to be with. They drop everything to go and see it. It is as though precious things cannot stay. Precious things seem to flee.

Okay, so precious things flee. What sorts of things flee? Flowers. Grass. People. Living things. The natural world flees (Isa. 40:6; Marchant, 1646; Wordsworth, 1807). Maybe that's why each of these things is precious? Yes, people can pluck them by their roots and press them in their diaries; but, like *the overview effect* from space (an extreme moment of clarity), they've preserved a great sorrow in exchange for its preservation (De Luce, 2019; Shaw, 2017). So many things people fear to lose are fleeting too, and so perhaps because people hope they will never lose them, they flee that much quicker. That first date you wished would never end, that last day with your dying mom or dad—it's almost as though one drives those moments away by clutching them so desperately and wishing they would never end. Frankl called this *paradoxical intention*—when Life seems to avoid whatever one wills it to do (Frankl, 2006).

Now try this in reverse. What doesn't stay precious? Better said, what *doesn't* flee? Cities do not flee. Cities are made of stone, not grass. It's hard to miss a city that is always there. It's hard *not* to miss the grass when all the grass is gone (and all that remains is a city). Mountains do not flee. Mountains take a long, long time to wither away, just like concrete cities.

So, if people teach their children to love a big city instead of the wilderness all around it, the children won't miss the wilderness when it's gone (or the city either)—any more than one might worry whether one's holy mountain will be there when one gets back. One needn't remember a thing that never leaves. One remembers what flees (Matt. 26:11).

How about something bigger than a city but not as big as the planet? Empires—human empires do not flee. In fact, empires usually stay around too long—so long that everybody *wishes* they would flee already. Empires endure for millennia sometimes; the thing about things that last too long is that they cease to be precious and begin to be the opposite—they demand to endure and in so doing, demand to no longer be involuntarily precious (Leiss, 1972). So, if one learns to love a huge, flourishing empire in preschool, one probably isn't going to worry about waking up someday and missing that empire. Yet they do. But they don't miss what their empire made extinct.

Okay, so precious things are transitory, and things that seek immortality are anything but precious to those whose entire existence is transitory. What are some more enduring things then? Before one answers that, isn't it odd that primarily non-living things endure? Isn't it odd that only Life is transitory? Viktor Frankl once wrote that the very nature and value of human existence (as mortal humans) hinges on the unavoidable fact of Life's transitoriness, and that knowing this brief time here was not in vain matters more than anything else (Frankl, 1990). How will others remember you after you're gone? Did you matter?

Now that these conjectures have connected the idea of transitoriness with the urgency to matter, it's time to ask again about their opposites. What are some things that civilizations craft that seek immortality and thus cease to matter at all (since everything that matters is brief)? Big stone buildings do not flee—the whole idea is for a structure to outlive the civilization that built it; big stone idols do not flee either—the whole idea of idolatry is to cling to something that outlives everybody, resulting in a focus on it instead of each other (Meinecke, 2018a, 2018b). Okay, so man-made works endure but eerily outlive their preciousness; what about living things? Do any living things endure? Well, no—a living individual typically does not last very long. However, people have invented a thing called a "species" to fix that problem, the same way people invented stone idols that look like people and animals (to outlive people and animals). When one looks, one sees that a "species" seems to endure (even if all the individuals perish all the time). A human being, by similar, is like a species as well. It seems to endure even though its trillions of cells are perishing all the time (Gilbert, Sapp, & Tauber,

2012; Meinecke, 2018a, 2018b). What is it about a species (or a life story for that matter) that makes this possible? According to evolutionary theorists, a successful species is one that outlives every other species by seizing and keeping control. Remember the idea of an empire outliving what mattered? A dominant species is like a dominant empire—it always outlives the diverse individuals which comprise that species or that empire, so that only *the idea* of its superiority to individuals survives. Like dominant empires, dominant species do not flee (and every other species wishes they would flee already, so the rest of the world can enjoy a bit of freedom for a change). In times past, some pretty nasty things were the dominant species (for example dinosaurs) and everybody else had to eke out an existence between their footprint. Humans are the dominant species today—people are like the new tyrant lizards that rule over the Earth with abject derision for anything less human than they—and most of Creation (along with the few remaining reptiles) has probably had enough of them (Costello & Hodson, 2014; Gibbons et al., 2000).

What else? As it turns out, the idea of an enduring species continues in the idea of an enduring human self. The lasting self and its mental story outlive its trillions of cells, just like the idea of an animal species outlives every single member of that species. Reason suggests that the sensible idea of a thing long extinct is surely more painful and less desirable a survivor, than an insensible thing not yet an idea (drawing from Darwin's conclusion in 1876 that each creature is part of an insensible series or kind). The singular idea of a species dominates and displaces the amazing diversity of its briefer members, much as the idea of a person reduces its myriad aspects to a simple label. In the same way, the singular idea of a self dominates and displaces the plurality of its briefer cells and instants, much as the idea of a species reduces its myriad members to a simple label (Gilbert et al., 2012). Nothing else remains, as the poem goes, but the shattered visage of the former fusiform idol (Shelley, 1817).

While the authors are reciting poetry, perhaps this section might reach into some more famous metaphors from English literature to help elucidate the human species' fixation on immortality. Housman wrote, "and the name died before the man" in *To an Athlete Dying Young* (Housman, 1896, 20). This fits what Frankl (1990) wrote about pretty well (*Facing the Transitoriness of Human Existence*). Why do humans want the *name* of a person to outlive his or her nameless instants? Why do humans want the *name* of a species to outlive each nameless creature that seemed part of that species? It's the same process; no matter how different people are from moment-to-moment, one overarching act impersonates all the wonderful scenes which were briefly a person. The Bard wrote "A rose by any other name would smell as sweet" (Shakespeare, 2.2.48-49, as cited in Sugg, 2016). This is also true of people—whether one is lastingly remembered as Sam or Sally, everything you more briefly were remains more unique than similar—even if what people call you is always the same. Is a man (the person to whom his name refers), not more important than his name? But having heard his name, having known him by name, long after he is turned to dust, others think they know him.

Humans like to think that they are irreplaceable . . . in relationships, in their jobs at work, etc. Maybe they are? Maybe a "job" that anybody who is qualified to fill can fill, has become more important than the individuals who may or may not be qualified for that job? Maybe human jobs (like human names), diminish the need for people to stay together so as never to miss a moment of one another's uniqueness. A skill only allows one to be lumped into a category with others that have the same skill. A label allows people who have never met you to act as though they know you. Baudrillard wrote extensively about this. His work is entitled, *Simulation and Simulacra* (Baudrillard & Glaser, 2014). In exchange for the ability to brag about what they "do" for a living, people waive the right to matter whether they "do" anything or not. This topic is called *ableism*; many children agitate for the right to matter, but will never matter because they lack the ability to work (Meinecke, 2017).

But Housman helps one see that "people" do not care about "the man" at all. People only care about his lasting fame, since "Today, the road all runners come" is the very last day of everybody's life. Every life ends in physical death—survived only by the memory of one's name (Housman, 1896, 5). It is like the building of an empire made of painful memories of what those awful people did, and it will outlive them and make their children sad long after they die. It is like an imperishable reign of personal bests and world records that others will envy and feel inferior to whenever they think about them. So why do people want their name to outlive then, if all it does is engender lasting sorrow or lasting envy or lasting shame for them (by dying too soon), or in those who outlive them (by living too long)? That is an empire of the self, a disembodied species that outlives each instant of its precious mortal time here. There is a song by Queen that captures this idea. It is entitled, *Who Wants to Live Forever*, and it echoes the angst of the survivor of a love affair that died too soon (May, 1986).

Human ideas do not flee, nor do human laws or human institutions flee (in fact, they demand to stick around, even if everything else perishes in the process of making sure the mighty works of Man never perish). When one thinks about it, nothing which is not deemed at risk of extinction is at risk of extinction. Therefore, whatever is not on the endangered species list (not at risk of extinction) is probably not deemed very precious either. Curiously, the human species is not on that list—yet it deems itself more precious than anything on that list. If Man's ideas do not perish (flee) though the world around them is perishing, what flees then? The authors are glad you asked that.

WHAT CANNOT STAY (Qui Fuit)

Long ago, someone named Jacques Marchant (1646) asked himself, *Qui fuit*? He thought and thought. He realized that childhood flees (*fugit pueritia*). This certainly rings true today. Childhood does flee, and all too soon people wish they could have remained children just a little longer. Marchant apparently thought some more—and realized that every other age of the human lifespan flees as well. He wrote elegantly that: infancy flees, childhood flees, adolescence flees, youth flees, and even that final part of life—old age (that seems to go by too slowly)—flees as well (*fugit infantia, fugit pueritia, fugit adolescentia, fugit juvenius, fugit senectus*; Marchant, 1646). Every epoch of the human lifespan worth remembering flees before people notice it; and it is when people lose those beloved moments completely, that those moments finally become worth remembering (Meinecke, 2017). In his profound words, nothing here today is missed quite as much as yesterday. Since yesterday flees, all those yesterdays linger for a second chance at today.

Again, what flees? According to Marchant, everything except the idea of God seems to flee (because the idea of God is really the need to get everything back again; Exline, Park, Smyth, & Carey, 2011; Gorsuch & Smith, 1983). Marchant had realized that for something to last forever, it has to end, so that a person can actually long to have it back, the same way a person longed for it to happen. Some have compared this to the need for one thing to give its life—so that many lives may be enriched from that moment on (John 12: 24-25). Memories certainly do seem to be the result of the loss of just one life, so that many memories may live on. Following that premature departure of the beloved moment, inside each grieving survivor are spawned many overdue thoughts, that continue to flower like determined lilies atop the grave they are tearfully left upon (Manning, 2011; Meinecke, 2018e). It might be said that people do not even notice what they treasure until they lose it. Childhood is such a thing. Like flowers, childhood is brief. As Doty (1984) put it so poignantly:

In a neighbor girl's front yard we played
the same game every summer twilight,
blue as Egyptian porcelain, that stretched
between supper and time to come in.
Every night we took turns dying.
One would lie down while the other
folded the corpse's hands and,
with the true solemnity of children,
brought flowers. (p. 118)

WHAT CANNOT STAY? (CHILDHOOD)

The fundamental paradox that arises whenever humans need to decide what to treasure most, is the great difficulty of distinguishing the unique value of many similar things—after going to so much trouble to look at all of them as similar (Meinecke, 2017). In the final analysis, humans wish someone else would decide for them (Roeland et al., 2014). If that is so, why do humans try so hard to decide the "right thing to do" in the first place? In the author's dissertation, it was found that the flourishing of an *intentional* process (rather than an unintentional process) depends largely on being able to reduce the pool of legitimate candidates for public regard to a heap of unremarkably perfect people. This perspective, as it turns out, has been deeply analyzed by the social cognitive lab at Princeton, and the model is called the *Stereotype Content Model* (Fiske, 2010). In fact, a prominent article by the principle investigator of this model described division by status as one or more *intentions* (the need to rank everybody from best to worst, even if everybody is different). In the author's dissertation, this model was found to encompass far more than just human attitudes toward other humans. This ego defense in response to other egos seemed evident whether the comparison was human versus animal (different species), more human versus less human (different belief systems), or more mature versus less mature (different ages; Freud, 1966; Meinecke, 2017). In this section, the authors will briefly investigate the phenomenon of a universal societal prejudice against human children. It appears to be an inadmissible envy rather than a justifiable disregard.

WHAT CANNNOT STAY? (MEMORIES)

Regardless of how Society regards its contents, not everybody looks back on the stages of their lifespan the same way Society would like everybody to look at them—for example, *too young* to work or *too old* to work. The regard Society has for these marginalized groups looks much more like disregard than "protection," much like the way Society regards its natural resources. For example, some think back on their childhood very fondly—a time of life when humans were once free of responsibility, and finding joy before the sun went down was the one and only long-term goal of every human child. Whether that child was too physically immature or too mentally immature made no difference to the briefer child, where his or her hope of transitory happiness was concerned.

Not all think of childhood quite so wistfully—in fact, they may prefer not to look back at all. Why? Perhaps, they were among so many hapless children who had *authoritarian* parents instead of plain old "I just love you as you are" parents. And so, those times beyond the grace of childhood, when grace is sparse and one needs to reflect on better days, one has no better days to reflect upon—save getting away from one's abusive parents so one can be oneself for a while (Moilanen, Rasmussen, & Padilla-Walker, 2015). Or perhaps, rather than being psychologically abused by inordinate expectations of perfection, they were physically abused (and they couldn't fight back due to their "age"). Either way, they probably could not *wait* to grow up… so they could move out of there—and escape their rules, their axioms, and their endless manipulations! And so, the very thing most long to get back to, was the very thing they longed to get away from. Less candidly, some have been the childhood victim of secret abuses rather than open abuses, acts protected by the silence of unrepresented children (Daniels, 2015). No matter how one looks back, when a human is "just a child", it has no right (either way) to represent itself, and must hope that at least one adult has pity on childhood. (This is also the case with Nature). Nor do the protected have the freedom to run away from those who "protect them" (be those self-ordained authorities legitimate or illegitimate in their own perception). This plight of human children is eerily like the plight of non-human animals too.

Now for an eye-opener. Maybe the period of the lifespan called childhood shouldn't be looked upon as part of a larger span? Maybe childhood isn't a part of anything? Maybe childhood is a species all by itself? And just maybe that's the whole problem? The authors ask that because, when a thing can only be one thing (a child, for instance), one doesn't have to divide it into smaller pieces to explain why sometimes it's a good thing to be a child and sometimes it isn't (or an adult, or a senior citizen). Maybe the whole reason humans see things so differently than the animals (even when science says both are animals), is because only *homo sapiens* does not accept a thing the way it is? A thing is always becoming a thing, or has been that thing and is now becoming something else. This is insensible in a troubling way, not in a delightful way—as Darwin (1876) envisioned no end to the species of the blue planet, third from the sun. If one eliminates these stages across the lifespan, one eliminates the conflict that arises by comparing each stage with every other stage, and no being at any stage can fail, because there is nothing to overcome (Meinecke, 2017).

WHAT CANNOT STAY? (AWE)

As discussed in chapter 2, the benefits of taxonomy exact an odd, unrecoverable cost equivalent to that benefit. Just as it is odd to heap millions of diverse people into one category (so humans can highlight the individual differences of each and every person), so too it seems odd to heap millions of diverse hours into one category (so humans can highlight the individual differences of each and every hour). Perhaps the very idea of human judgment arises because humans heap many different things under one umbrella concept, only to point out how much better or worse each thing is relative to every other thing in that umbrella category. This was the gist of one of the conclusions of the author's dissertation (Meinecke, 2017). Maybe (just maybe) humans do not need to compare one another to some ideal? Interestingly, President Theodore Roosevelt seems to have said the same thing:

Comparison is the thief of joy

In order to fully understand the concept the authors call *biophilia*, it seems helpful to cast away assumptions that might prevent humans from being amazed at the sheer wonder of this planet full of living people and living things in general. One of those invariant assumptions in need of challenge, is that people share certain discrete traits in common (rather than each individual being remarkable for its own unique traits). The second assumption in need of challenge is a lot like the first—it is the assumption that every Life consists of discrete stages (rather than the possibility that each part of the lifespan is a new creature in its own right). Under the common assumption, there is only one ideal segment of the human lifespan which is looked upon as everybody's purpose for being here (while every other segment of the lifespan is perceived as preceding or following that perfect period of life). In this author's dissertation (Meinecke, 2017), it was proposed that *the age of industry* is seen as the purpose for human existence (by their societies). For example, before the age of industry, humans are *too young* to be of much use. And after the age of industry, humans are *too old* to be of much use. Most sparse resources seem to be devoted to that one stage when human beings are the most "productive" (not fertile in the ancient sense and able to produce offspring—productive in the sense of churning out lifeless products to be "consumed" by the working public). In the eyes of each society, only one stage of this approximately 100-year lifespan of ours is *just right*. The rest is just wasted apparently. That period just happens to coincide with when humans are employable (consumers of human products) by the society humans happen to have been born into (usually 18-64 in the U.S. these days, although the ideal age group varies by region and by time period). Outside this ideal period, humans are not considered fully human. In fact, if humans do not pass certain criteria, they may die before they are considered human—even if they were obviously very much alive.

Second, even if one's *physical age* is within the sacred parameters of one's society, nowadays one's *mental age* must also fall between the boundaries human society has invented as criteria. Physical age is fairly evident; it is pretty hard to hide how young or old you are. Physical bodies all grow old and die. Mental age is different though; it is hard to tell a person's "mental age" from afar. How tall must one be to be intelligent? Some child protégés are far more intelligent than grown-ups three times their physical age. And many old people are far wiser than Society gives them credit for, simply because they are no longer employed. Because there are lots and lots of people who fit in the ideal physical age category, humans find themselves using an odd criterion called IQ (a sort of mental equivalent of physical age) to narrow this down to a subset of perfect humans. This is rationalized by the complicated regimen which must be understood and followed to survive in a civilized world.

Third, even if a person looks physically the right age, or mentally the right age, he or she may not be as "able" as others (Meinecke, 2017). Psychology has words for this: *ageism* discriminates against those too young or too old to have the same rights as everybody else. *Ableism* discriminates against those who are seen as being less fit than the average person of a given physical or mental age.

Does the reader see a pattern here? How can a person know when he or she is human? Is a child fully human? When is a child human? Human children do not have control over their own lives—they are seen as the protected property of grown-ups, under a perspective often framed as endearing yet inferior (in paternal societies). Not so long ago, American women were also viewed like children (endearing yet inferior) by their paternal societies (as were pets). In that age scarcely a hundred years ago, gender was the criterion instead of age or intelligence (and in many ways, it still is).

So what does it mean to be in love with life (reverence for life)? What does it mean to be in love with living things (biophilia)? Must they be a certain age to qualify to receive societal reverence and love? Must they be a certain gender? Must they come from or practice a certain "way of life"? More tragically, must humanity view its childhood as preparatory to an *Uberzeit* called middle age? Must the elderly view their golden years as an *Unterzeit* (compared to younger, "working" adults). Must everybody who isn't at work or wishing they weren't at work, look wistfully back on the Life that is passing everybody by? The authors wonder, is this rational?

But what if humans did not divide the human lifespan into stages? Who says humans have to? Who told humans they were naked? (Gen. 3:11). Isn't it curious that these bizarre problems go away when humans don't think they need clothing just to exist, or divide their moments into too many moments, not enough moments, too good, not good enough, and so on? What if all that science asked itself before it went ahead and did something to another thing without its permission was, "Is it alive?" That should be the sole condition for reverence for Life (not consciousness, or vertebrae, or intelligence, or gender, or age, or race, or way of life, or some other handy reason to deny a living thing the right to simply live). Be that lifeform connotative of *infantia* (infancy) or be that lifeform connotative of old age (*senectus*), the operant morpheme is "life" not form. Otherwise, some will look back on their childhood with sorrow mixed with fear (*tristitia* and *timor*)—thankful only because they are no longer rightless children. And others will look back on their childhood with sorrow and loss (*tristitia* and *damnum*)—desperately wishing they could simply stop working (to earn things they may never survive to enjoy). Almost four hundred years ago, someone in academia, but a much more reverent academia, asked the very same question. If everything flees, then, does anything remain? (*Quid est Deus*?). Yes. Something always remains—the one thing that cannot leave, if everyone is to realize the value of what flees (Matt. 26:11).

HONOR FOR THE LEAST OF THESE

It seems to follow from the above arguments, that it is unwise of this species to put the majority of a thing first for survival and regard, when there are so many of it, and they are already fortunate to be among that majority, and each of them is so easily replaced (according to human criteria anyway; Meinecke, 2017). It seems antithetical to hold that even though humans are unlike what is precious, unlike what is rare, and unlike what is brief, humans are so precious and rare and brief that they would gladly let the entire world perish if they couldn't appreciate and control it (Leiss, 1972). It makes so much more sense to think outside the box and wonder whether the least common among the most common thing is the "wildest, the most urgent" member of their kind—and, therefore, the most passionate, the most alive among that set of unremarkably conforming similars (Lawrence, 1920). When the reader practices that simple thought experiment, something changes inside; one is almost "transformed by the renewing of your mind"; one suddenly realizes that all of it is lovely, if the least comely part is lovely (Matt. 25:40; Rom. 12: 2-3; Phil. 4:8).

Suddenly the awakened heart within notices that every species is worth saving. Now, one no longer has to consider "which member" of a species is most like that species or "which set of species" is most representative of all the other fauna and flora on the planet, and therefore worth saving (they all are). And, most curiously, one begins to look at the brief moments one has on this Earth in the same way, and at both one's neighbors and one's "enemies" with a new heart (Costello, 2013; Costello & Hodson, 2014; Matt. 5:43-48; Psalms 51:10).

It follows then, that if the human species should try and see animals with more reverence than it typically does (and this results in more reverence for human outgroups) then if humans should try and see *flowers* with all the reverence they can muster, how much more might they revere each other? (Costello & Hodson, 2014; Luke 6:37; Luke 12: 22-31). If you practice reverence for the least of these, the greatest of these will also know just how much *more* you revere her—even if that lesser thing is as simple as a rose, and that greater thing as amazing as a beautiful woman.

CONCLUSION

The title of this chapter was rather odd, wasn't it? How could reverence for a flower save a dying planet? Can flowers do miracles? Can plants save the planet? Why should civilized people even care about flowers? They don't even have a brain. Well, to be honest, it is very unlikely that making daily offerings to wildflowers will elicit their goodwill, and Nature will suddenly bloom in kind. However, the transformation of human *attitudes* could very well make all the difference (Costello & Hodson, 2014). Practicing a simple gesture like kneeling to an incidental flower along the way whenever it is stumbled upon, could spark in each individual the same stimulation to kneel to each other just to say how grateful each feels to see the other each day. As any psychologist will tell you, if you are not reluctant to do a hard thing, you will find it a lot easier to do an easy one. Seeing a flower with reverence is a hard thing; if nothing else, your friends will look at you funny (and might refer you to a psychologist!). But if humans can become accustomed to doing this little thing easily, it should not be very hard to stop along the way during their busy-busy lives, and just adore one another openly each time they get the chance. Too often human goals or the unfamiliarity with that sort of uninhibited reverence holds humans back (Darley & Batson, 1973).

This chapter was about challenging human thoughts about what matters most. Do humans matter more than the planet they walk on? Whatever is precious should come first. Whatever is precious should matter most. What is more precious than Life of every kind? In these pages, what was learned was that whatever is the most rare in all the universe remains infinitely precious—no matter where one happens to be with respect to it. It was also learned that what cannot stay (especially when nobody can imagine losing it) remains precious yesterday, today, and always—which kind of unchanging affection in Latin is usually rendered *immutabilis* (Marchant, 1646). Whatever is precious is first, is rare, and cannot stay; and because of these three things, it never seems to go away, even though it flees.

Why do humans refuse to go away? Who do human jobs come first? Once, long ago, humans awoke to a sunrise in which every creature shared the same sunlight under the same sky. Now humans mask the billions of nocturnal suns so a few artificial lights from these man-made suns can light the night as well as the day, and compel billions of bipedal oxen (yoked to their desks) to keep pulling their ploughs, long after the starry fields have drifted off to sleep. Nature has no place left to sleep. Humans have no refugia either, but only Nature realizes it.

But there is a bright new star over a manger; perhaps a simple change of reverence for a child or a minor thing like a rose is long overdue; perhaps, if humans could only kneel to a flower as though to a coming king, that kingdom might come.

REFERENCES

Baudrillard, J., & Glaser, S. F. (2014). *Simulacra and simulation*. Ann Arbor: The University of Michigan Press.

Costello, K. (2013). *Determinants and consequences of dehumanization: An interspecies model of prejudice* (Doctoral dissertation). Brock University, St. Catherines, Ontario, Canada.

Costello, K., & Hodson, G. (2014). Explaining dehumanization among children: The interspecies model of prejudice. *British Journal of Social Psychology, 53*(1), 175–197. doi:10.1111/bjso.12016 PMID:23083220

Daniels, D. G. (2015, May 15). The race of my life: A runaway kid's story. *Children's Rights*. Retrieved from https://www.childrensrights.org/the-race-of-my-life-a-runaway-kids-story/

Darley, J. M., & Batson, C. D. (1973). From Jerusalem to Jericho: A study of situational and dispositional variables in helping behavior. *Journal of Personality and Social Psychology, 27*(1), 100–108. doi:10.1037/h0034449

Darwin, C. R. (1876). *The origin of species by means of natural selection, or the preservation of favoured races in the struggle for life* (6th ed.). London, UK: John Murray.

De Luce, I. (2019, July 16). Something profound happens when astronauts see Earth from space for the first time. *Business Insider*. Retrieved from https://www.businessinsider.com/overview-effect-nasa-apollo8-perspective-awareness-space-2015-8

Deleuze, G., Guattari, F., & Foucault, M. (2009). *Anti-Oedipus: Capitalism and schizophrenia*. New York: Penguin.

Doty, M. (1984). A replica of the Parthenon. *Ploughshares, 10*(4), 118-120. Retrieved from https://www.jstor.org/stable/40349318

Exline, J. J., Park, C. L., Smyth, J. M., & Carey, M. P. (2011). Anger toward God: Social-cognitive predictors, prevalence, and links with adjustment to bereavement and cancer. *Journal of Personality and Social Psychology*, *100*(1), 129–148. doi:10.1037/a0021716 PMID:21219078

Fine Gardening. (2020). *Darcey Bussell rose*. Retrieved on March 10, 2020 from https://www.finegardening.com/plant/darcey-bussell-rose

Fiske, S. T. (2010). Envy up, scorn down: How comparison divides us. *The American Psychologist*, *65*(8), 698–706. doi:10.1037/0003-066X.65.8.698 PMID:21058760

Frankl, V. (1990). Facing the transitoriness of human existence. *Generations (San Francisco, Calif.)*, *14*(4), 7. Retrieved from https://www.asaging.org/generations-journal-american-society-aging

Frankl, V. E. (2006). *Man's search for meaning*. Boston, MA: Beacon Press.

Freud, A. (1966). *The ego and the mechanisms of defence*. New York: International Universities Press.

Fromm, E. (2010). *The heart of man*. Riverdale, NY: American Mental Health Foundation Books.

Fromm, E. (2013). *Escape from freedom*. New York, NY: Open Road Media.

Frosch, W. A. (1970). Psychoanalytic evaluation of addiction and habituation. *Journal of the American Psychoanalytic Association*, *18*(1), 209–218. doi:10.1177/000306517001800111 PMID:5451017

Gibbons, J. W., Scott, D. E., Ryan, T. J., Buhlmann, K. A., Tuberville, T. D., Metts, B. S., ... Winne, C. T. (2000). The global decline of reptiles, déjà vu amphibians. *Bioscience*, *50*(8), 653–666. doi:10.1641/0006-3568(2000)050[0653:TGDORD]2.0.CO;2

Gilbert, S. F., Sapp, J., & Tauber, A. I. (2012). A symbiotic view of life: We have never been individuals. *The Quarterly Review of Biology*, *87*(4), 325–341. doi:10.1086/668166 PMID:23397797

Gorsuch, R. L., & Smith, C. S. (1983). Attributions of responsibility to God: An interaction of religious beliefs and outcomes. *Journal for the Scientific Study of Religion*, *22*(4), 340. doi:10.2307/1385772

Haugeland, J. (2013). *Dasein disclosed*. Cambridge, MA: Harvard University Press. doi:10.4159/harvard.9780674074590

Housman, A. E. (1896). *To an athlete dying young*. Retrieved from https://www.poetryfoundation.org/poems/46452/to-an-athlete-dying-young

Knickmeyer, E. (2019, August 12). *US government weakens application of Endangered Species Act*. Associated Press. Retrieved from https://www.apnews.com/9bf4541d89e6444783814e53302ce479

Lawrence, D. H. (1920). *Whales weep not*. Retrieved from https://poets.org/poem/whales-weep-not

Leiss, W. (1972). *The domination of nature*. New York, NY: George Braziller.

Lenggenhager, B., Tadi, T., Metzinger, T., & Blanke, O. (2007). Video ergo sum: Manipulating bodily self-consciousness. *Science, 317*(5841), 1096–1099. doi:10.1126cience.1143439 PMID:17717189

Manning, D. (2011). *Don't take my grief away from me*. Oklahoma City, OK: InsightBooks.

Marchant, J. (1646). *Rationale evangelizantium, sive doctrina et veritas evangelica, a sacerdotibus, pastoribus, concionatoribus, pectori appendenda, plebique per anni circulum è Cathedris proponenda, exponenda: Duobus tomis comprehensa*. Parisiis: Apud Michaelem Soly.

May, B. (1986). Who wants to live forever [Recorded by Queen]. On A Kind of Magic. Munich, Germany: Musicland Studios.

Meinecke, L. D. (2017). *Neglected by assessment: Industry versus inferiority in the competition for scarce kidneys* (Doctoral dissertation). Available from ProQuest Dissertations and Theses database. (ProQuest No. 10689852)

Meinecke, L. D. (2018a). A theory of cognitive idolatry, or the human struggle to preserve psychological certainty in a biological species - chapter 1 (Research proposal). doi:10.13140/RG.2.2.16527.43684

Meinecke, L. D. (2018b). A theory of cognitive idolatry, or the human struggle to preserve psychological certainty in a biological species - chapter 2 (Research proposal). doi:10.13140/RG.2.2.31089.53601

Meinecke, L. D. (2018c). *Hide her under a bushel? No, I'm gonna let her shine...* Retrieved from https://www.linkedin.com/pulse/hide-her-under-bushel-im-gonna-let-shine-lonny-meinecke-phd/

Meinecke, L. D. (2018e). *The uncanny fear of loss, part 1: Losing the unthinkable*. Retrieved from https://www.psychologytoday.com/us/blog/theory-and-praxis/201804/the-uncanny-fear-loss-part-1

Meinecke, L. D. (2020, March 7). *Nature-nurture and language development*. Retrieved from psyarxiv.com/4u3kq

Milton, J. (1650). *Sonnet 19: When I consider how my light is spent*. Retrieved on March 10, 2020 from https://www.poetryfoundation.org/poems/44750/sonnet-19-when-i-consider-how-my-light-is-spent

Moilanen, K. L., Rasmussen, K. E., & Padilla-Walker, L. M. (2015). Bidirectional associations between self-regulation and parenting styles in early adolescence. *Journal of Research on Adolescence*, *25*(2), 246–262. doi:10.1111/jora.12125

Pais, A. (1979). Einstein and the quantum theory. *Reviews of Modern Physics*, *51*(4), 863–914. doi:10.1103/RevModPhys.51.863

Roeland, E., Cain, J., Onderdonk, C., Kerr, K., Mitchell, W., & Thornberry, K. (2014). When open-ended questions don't work: The role of palliative paternalism in difficult medical decisions. *Journal of Palliative Medicine*, *17*(4), 415–420. doi:10.1089/jpm.2013.0408 PMID:24588626

Rogers, C. R. (1961). *On becoming a person*. Boston, MA: Houghton Mifflin.

Sagan, C. (2019). *Carl Sagan: Quotable quote*. https://www.goodreads.com/quotes/144310-the-nitrogen-in-our-dna-the-calcium-in-our-teeth

Saint-Exupéry, A. (1943). *The little prince*. New York, NY: Harcourt, Brace & World.

Scheele, D., Wille, A., Kendrick, K. M., Stoffel-Wagner, B., Becker, B., Güntürkün, O., ... Hurlemann, R. (2013). Oxytocin enhances brain reward system responses in men viewing the face of their female partner. *Proceedings of the National Academy of Sciences of the United States of America*, *110*(50), 20308–20313. doi:10.1073/pnas.1314190110 PMID:24277856

Shaw, S. (2017, January 17). The overview effect. *Psychology in Action.org*. Retrieved from https://www.psychologyinaction.org/psychology-in-action-1/2017/01/01/the-overview-effect

Shelley, P. B. (1817). Ozymandius. In *Shelley's poetry and prose*. New York, NY: W. W. Norton & Co.

Sugg, R. (2016). Shakespeare's anatomies of death. In M. Neill & D. Schalkwyk (Eds.), *The Oxford Handbook of Shakespearean Tragedy* (pp. 184–201). Oxford: Oxford University Press.

Vrobel, S. (2011). *Fractal time: Why a watched kettle never boils*. River Edge, NJ: World Scientific & Imperial College Press. doi:10.1142/7659

Wells, H. G. (1898). *The war of the worlds*. Leipzig, Germany: Bernard Tauchnitz.

Wordsworth, W. (1807). *Ode on intimations of immortality from recollections of early childhood*. Retrieved from https://poets.org/poem/ode-intimations-immortality-recollections-early-childhood

ENDNOTE

1 Discrete infinity is from Rieber's (1983) interview with Noam Chomsky – his elegant term for human utterances.

Chapter 12
Beyond First Love:
No Greater Love

ABSTRACT

The theme of this final chapter (no greater love) is the simple concept that, if the human species really cherishes this beautiful world, it will gladly give anything to save it. Instead, this species puts itself first, and holds that the "commercial prosperity" of just one vain and merciless species is more important than the biological survival of every other species. But imagine a people who sacrificed their kind to save life itself from perishing. Isn't that a purpose worth setting aside global differences? Imagine perhaps the idea of no greater love is about to find a place in an imprescriptible history that illiterate kinds will benefit from forever. This species, more than any other species, may join together as one humble and reverent kind and make every effort toward saving this beautiful planet from its longitudinal yet incidental myopia.

INTRODUCTION

Greater love hath no man than this, that a man lay down his life for his friends. - Jesus (John 15:13, KJV)

This theme, "no greater love," is the simple concept that if humans really cherish this beautiful world, they will gladly give anything to save it. Nothing is more precious; nothing is of a higher priority. It is like the groom's promise

DOI: 10.4018/978-1-7998-4408-2.ch012

to his bride, to forsake all else because she is the one pearl of great price in all the world, and he would do anything for her. It is time now, a chance for humankind to be that groom. It is her time now, this beloved world's chance to shine and be the center of every attention. This is her Day.

Instead, the human species is like an unfaithful husband with a penchant for *narcissism* (Freud, 1957). The species is a just a self-propagating paternalism whose spouse (the planet) is no more than an endearing trophy reminiscent of his stature and success (Meinecke, 2018c). Humans are like a loveless parent with an *authoritarian parenting style* (Darling, 1999). Humans are more concerned with using their offspring to agitate for the right to compete with men and control the *means of reproduction*—than to express a simple gratitude and mercy for those offspring (Goldberg, 2009; Meinecke, 2017). It does not seem urgent any more to simply hope to pass down a memory of how much one could not help but love one's husband, one's wife, one's children, one's world, while briefly here—so that, perchance, they might pass down their love for one another too (John 13:34; Gopnik, 2016). Humans put their doctrine first; they put themselves first; they put their estates first; they do not seem inclined to ask their lovely bride if she would like to be *first* for a change. Humans are a species whose own "prosperity" and "wellness" is more dear to them than the well-being of their beloved world. they would rather put their species' greed first (the craving for global sovereignty and imperishable prosperity of just one species), and let that discrete infinity of basic needs requested by every other species (save their own), glean what means for group survival or personal value they may (Lev. 19:10; Rieber, 1983; Ruth 2:10). Human vows are like those of a dissembling bachelor at the honest altar of his virgin bride, withholding his ability to give himself fully to save his beloved wife, so that he may simply wed another when this one wears out (Steinbeck, 2003). Though humans have more wealth and more options than any other creature on the planet, they make every other creature sign a prenuptial covenant before they vow to demonstrate their love for them at all.

But imagine a People who were willing to sacrifice their own *kind* to save Life (in all its *kinds*) from perishing? Isn't that a purpose worth setting aside these differences? Imagine a love so great, that a simple act of faith might save the world from its victor's folly? Let all things great and small sit down and reason together: No greater love hath any species than to lay down its pride for its pearl of great price—this sapphire earring the grateful call Home (John 15:13; Isa. 1:18; Matt. 13:46). Think about it: it's the *only* home for every species living today (so far as anyone knows), not just the home of human beings. Though humans may build fleets of rockets to save themselves from the coming deluge of commodities and pollutants, no ark could begin to carry the diversity of lifeforms every bit as urgent to survive as humans—but unable to even build a life raft once humans depart, let alone a rocket to bear their young to safety (Bradbury, 2001). Humans would make room in their arks for their wisdom (an intellectual *posada*), but not for their faceless children (*no posada).* What does that say about the human species?

NO GREATER LOVE

This final chapter in this book about biophilia is about how to save a planet from dying, because no book about reverence for Life would be complete without a proposition of how to become more reverent (Matt. 19:16-26). In the last chapter, the authors talked about a change of value systems—how reverence for a flower could renew the first love humans had for their world. But up till now the authors have only spoken of what biophilia is and isn't (biophilia is like unconditional love, not jealous protectiveness), or about societal indifference to the imminent extinction of all life on this planet. Now this text will become practical—now the authors will propose a tangible way to save this dying planet from certain death. To paraphrase a famous speech by J. F. Kennedy: we choose to do this not because it is easy, but because it is (so) hard (NASA, n.d.). We choose to do this because nothing matters more than a chance for an undiscovered country to find a place in the sun, and greet the coming of a new Day with gratitude rather than fear (Hansberry, 1959). We choose to offer up ourselves as living sacrifices to save our one and only Bride, this lovely thing we call *terra* (Mitchell & Conroy, 2011; Rom. 12;1-2).

Will this work? Let the authors put it this way; if the one thing humans will not do is risk their own extinction, then the willingness to risk their extinction because they revere Life of every kind more than they revere ourselves, is one more chance to save their planet from certain extinction than they now have. Perhaps that was how the wizard of Menlo Park (Edison) approached each impossible task? The one way that will work is the one way humans did not try. There is a linguistic concept called *clusivity*. It is a verb form that is neither an "I" nor does it ignore the idea of a "We". Rather, it is a selfless we, the inclusion of all things one identifies with except oneself (O'Grady & Meinecke, 2015). The authors propose the use of this nounless noun as a selfless pronoun, as humanity approaches the struggle to save the world from extinction, by trying to perceive its existence as a chance to save everyone else's existence. As the primary author of this work so elegantly phrased it, ". . . all species both non-human and human are inextricably linked . . ." (O'Grady, 2016).

In any conflict, the willingness to do what nobody else will do can be the winning advantage; however, to date the willingness of most has only been to do some *awful* thing nobody else is willing to do. What if the winning advantage were the willingness to do that one *good* thing nobody else is willing to do? What then? Would that help? A species that is willing to leave its own interests out is able to contribute more than any other; this generation can be that People—a people such as the world has never seen, and will probably never see again—but the Earth will survive, because humans laid down their lives for their less human friends.

WHAT LETS OTHERS DIE IN ITS STEAD?

The second and only other point the authors wish to make for this final chapter's theme, is really the antithesis of *no greater love*. Sadly, more abundant than things that give themselves selflessly for others, are those that selfishly demand that others die in their place. What normally survives the general conflict is oddly more narcissistic and banal than what came before; what conquers is (nearly always) less kind than the thing it conquered (Cohrs, Petzel, & Funke, 2012; Jetten & Mols, 2014; Newlands, 2014; Newman, 2014). Human concepts tend to survive, but those who gave their all to save their concepts from perishing, never survive (Meinecke, 2018a, 2018b). Yet, since what survives is all that remains to record the difference between good and evil, no record is kept of what ought not to have survived, and what remains tends to believe in its inherent goodness (Irvine, 2019). Perhaps, then, it is this one thing that prevents the saving of a selfless planet? If humans have faith the world will be okay without them, perhaps the planting of that tiny seed is all the Earth needs to begin again.

This proposal of species hominin's potential selflessness is not new solely because of its novelty. The very model of survival among humankind is the idea that most should die—so that a few might live (or that some single ideal might live, more typically). Many term this *survival of the fittest*. But if one does that, those that survive will then be the fittest yet least diverse, benefitting from the plural sacrifices of the infirm among their kind, and the unblemished among every other kind. The unfortunate offset of this design is the preservation of egocentrism in the genes of the species, and the systematic culling of those deemed unrepresentative of the presumed superiority of the human species (Keltner, 2009).

But the human species is *not* special; the human species is *not* remarkable. The human species is remarkable primarily for its *belief* that it is special. Only this unflinching hubris in its species is a truly remarkable trait in an otherwise humble Nature, an arrogance that self-situates its kind among the gods and angels while placing every other species in Creation among the basest of beasts (Gilhus, 2006). This species holds that it is born imperfect, but by a stern and clever nurture is obsessively purified of its animal nature, and thus saved from its baser instincts—when the scientific evidence suggests the opposite is more in evidence (Keltner, 2009). The evidence so far is primarily that humans begin naïve and unaggressive and pure (much as any juvenile animal), but the systematic denaturing and reprogramming of its offspring compels the best survivors to become arrogant and self-sufficient, and likewise look down upon every other species (if they wish to be counted among the god-like human group and not be compared to the outgroup beasts; Meinecke, 2017). The principle trait that exhibits this scorn of biological phenomena is the human mind—a thing uncommon among most other kinds (Meinecke, 2018a, 2018b).

In this section, the authors would like to submit that the reason this proposal is so hard to ask (this offer to sacrifice the human species if it might somehow save the planet), is because humans views the world as here to sacrifice itself to save humans. Humans are the only species that swap their organs in and out to prolong their mental activity; the human species is the only species whose devotion is toward preserving their mental phenomena instead of the planet's physical phenomena. Humans sacrifice millions of animals every year to prolong their already overextended lifespans, and to increase their own health and wellness while destroying theirs (Wanderer, 2015). Humans abbreviate the lives of every other creature because they think they have a divine right to live forever—yet this suggests such irreverence for Life, such ingratitude for the lovely moments humans are given (Meinecke, 2018a, 2018b). How can this be a divine commission if it is so completely different from the divine example humans were given—that God so loved the world he died to save it, not that God so loved Himself he demanded the world should die in His place—to save one arrogant species that sits on the throne of Life? (1 John 3:16). And good science defuses the claims of this species that humans are unlike any other, challenging both reason and conscience—challenges it to humble itself and include all beings in the idea of a declaration of universal rights, not exploit the others because it claims the divine right to exploit them (Meinecke, 2017; United Nations, 1948).

Biophilia—the love of living things—begins with identification with those things, not a posture of mental superiority and identification with gods. It is an odd thing that where one does not see vanity and certainty, one frequently sees awe and wonder; where one sees incompetence and clumsiness around the Lovely, one sees a reverent guy trembling when he is near her (Meinecke, 2017). And where one sees an incapacity to categorize and aptly codify Beauty, one cannot help but see an endless reverence for Life, as though meeting the play of light and shadow in the forest and the shimmering sunlight on the water for the first time—day after day after day. Some men see the wilderness and see a waste of good territory (Matt. 26: 6-9). Others see that same wilderness and thank God for its freedom. The ability to save the world you love depends on your willingness to save the sparkling ripples in a pool from extinction, not your own reflection in that water from extinction (Grandin & Johnson, 2005). The option needed to have any chance of rescuing the Earth, requires not being conformed to this world's human view of itself as divinely selected, and letting slip these assumptions of greatness to permit an awe for this natural world that nothing is greater than (Rom. 12:2). It is like a new heart, a clean spirit, and a long overdue transfiguration; isn't it time for a refreshing, from the vain and egocentric mind of Man into one that cannot judge which creature is greater or less than himself, because he reveres them all *more* than himself (Isa. 28:12-18; Psalms 51:10; Rom. 12:2).

The current mindset is still that of a ruler who lets others die in his place, and would rather let the planet die than reduce his emissions. The authors propose that if humans simply change this attitude to one of humility, reverence, and gratitude, they will gladly risk all they are and all they have to save their beloved world from their own economic folly. Such an option has not been proposed that the authors know of; this option does not assume the human race must survive to see if the planet does, but as an act of faith promises to give its all in the belief it can do what it promises to do and does not need to survive to see the results of its sacrifices (Matt. 8:5-13).

RECAP OF SECTION I: EXAMING BIOPHILIA

This is the point at which to stop, reflect, and look fondly back on everything introduced so far, in this examination of biophilia and societal indifference to environmental protection. How did this book begin? What did it cover? Did it offer a means to save the planet? Is it doable? Or is human reluctance to care more about other creatures than they do about themselves greater than their willingness to do something wonderful? In chapter one, the authors proposed a novel theory:

Within the earth's biosphere, all species both non-human and human are inextricably linked and innately motivated to consistently focus on and to interact with other species. When these motivations to focus on and interact with other species demonstrate the characteristics of awe, reverence, respect, and/or empathy, these inter-species interests and interactions can be mutually beneficial psychologically, biologically, emotionally, and spiritually in ways that encourage the species' survival, evolution, development, and ability to flourish. Conversely, when these inter-species interests and interactions exhibit characteristics that are indicative of egocentrism, self-serving biases, devaluation, and domination of one species by or over another species, the outcomes can be catastrophic not only for that specific interspecies exchange but also across the entire biosphere resulting in the decimation, destruction, and/or extinction of both known and yet-to-be discovered species. (O'Grady, 2016)

The authors didn't know where to start, so they started at the beginning (which is a very good place to start, according to the famous song in *The Sound of Music*; Wise, 1965). This examination of biophilia began with a discussion of adoration of life and living things (biophilia) versus fear of life and living things (biophobia). The first section pointed out that the cries of a dying world continue to be denied and ignored by a species exclusively concerned with human-to-human relationships—even though humans historically and consistently represent only a small portion of life on the planet. It deconstructed the concept of biophilia into some factors that seem to be present whenever reverence for Life is present, including awe and empathy. One of the authors reached out to the Pope for affirmation that all living things are part of Creation and worthy of adoration—and received a positive reply. Will animals go to heaven? Who knows—but there is something humans can do right now to make heaven seem possible for them without having to die to get there.

The first book section went on to describe how biophilia has tangible benefits. Biophilia can help cure diseases if humans collaborate with living things instead of enslaving the weak and extirpating the strong. Living things that aren't particularly human in form can comfort the human species in its grief in ways that its own kind cannot (or will not). Perhaps it is not a good idea to exterminate humanity's hope of physiological comfort and companionship, and keep insisting on verbal self-persuasion and loneliness instead. The first section went on to emphasize that the idea of Life does not even end at animals, but continues in manifestations of living wonder and endless majesty that exceed the idea of humans and animals combined—in vibrant forests whose neglected paths can restore the human spirit, if only humans will walk them again. This first chapter shared that this form of therapy is called *shinrin-yoku* or forest-bathing. Somehow, being enveloped by living things in the canopy, treading lightly on the understory, and revering them rather than fearing them, renews in humanity a right, contrite spirit, and a cleaner, more consolable heart (Psalms 51:10-17).

The ancients, it was discovered, though lacking in modern technology and medicine, found in reverence for living things both ways to cope and ways to heal, and were not too proud to view the behavior of animals as something to look up to and symbolize their joy of life. The deep respect still practiced by indigenous peoples for indigenous territory reminds humans that they have lost their intimate connection with this living world in their struggle to subdue it, and perhaps it is this *first love* humans need to get back again (Gould, 2011).

In chapter two, having described what biophilia must feel like in one's own backyard on Earth, the examination of awe for Nature pondered whether this reverence might continue into the night sky. The principal author decided that it did continue:

The Theory of Astrophilia suggests that sentience as defined as a capacity to feel, perceive or experience subjectively, combined with awareness as defined as a recognition of what we are not, provide the means through which we are inextricably linked with, and innately motivated to seek out the "other" with whom we share the Universe. (O'Grady, n.d.)

Just as the universe seems to begin just above human thoughts (at the *macrocosmic* level), and just beneath human thoughts (at the *microcosmic* level), so too reverence for Life continues beyond the familiar—into Black Forests made of galaxies of gases and stars instead of shadowy woodlands made of trees and critters—where wolves in the form of extreme distance and extreme cold stand between Awe and the other side of Adoration. But the human relationship with Nature is not impeded nor diminished by the wiles of distance, only made more secure by a *wanderlust* this gravity and this atmosphere cannot keep this species harbored within for long. From the finitely familiar down here to the infinitely unfamiliar out there, the search for *Tier mit der Heimvalenz* (a companion animal whose *valence* reminds everyone of home) takes the human species to the stars (Deleuze & Guattari, 1987, p. 357). So the principal author further decided that:

The Theory of Astrophilia proposes that a similar relationship exists among all sentient beings that share a common universe. Even beyond the familiar spheres confined to the gravity of our own world, there lies a celestial sphere, one might say, with which all beings are inextricably linked, be they confined to gravity, or be they beyond its influence. (O'Grady, n.d.)

Gravity is a good word, because the longing for things just out of reach begins where one's own grasp ends. Must the human species confine itself to what is familiar before it deems them its companions, or might that species extend this love of Life to *lovers without boundaries*? Maybe the riskiest affection, the most rewarding affection, begins with *das Unheimliche* (the uncannily familiar) because the Other is when and where this specific love expands to a universal love, and this body expands to a Body without borders, neither one any longer subtended by time nor separated by space.

With the expansion of the idea of reverence for the environment, there logically follows an expansion of the idea of protecting that environment. If humans by their own lack of foresight have injured their world beyond repair, might they learn at last not to carry their prior colonial perspective beyond its sullied atmosphere? For when one's motives are reverent not condescending, that which was once ignoble begins a noble metamorphosis—and in lieu of a mandate for dominion there reigns a mandate for mercy. The new prime directive might be one of an indomitable awe, of non-interference rather than intelligently intervening upon everything less intellectual, and a conditionless extension of the Belmont Principles to encompass *every* group in danger of invasive study and modification, not just a few protected human groups. Perhaps the idea of environmental protection ought to begin with viewing this species' own bodies as fragile natural environments, and that whether for the proximal (these perishable bodies) or whether for the distal (those imperishable celestial bodies) there ought to extend an endless reverence for both—not the mandate to treat both as inferior to the mind that appreciates them. Astrophilia means to be in love with the Stars above, just as biophilia means to be in love with Life down here. Perhaps both spheres are simply one environment separated by this species' incapacity for humility? For whether it is with regard to this biological temple, or whether it is with respect to some cosmic temple, it is hard to be humble and sovereign at the same time. It is difficult to gape "in awesome wonder" while sitting in judgment of every mountain and every valley from the event horizon of universal regard. What kind of ethics might humans prescribe that could guarantee reverence for all creatures while retaining this ingroup favoritism? The authors ask: what may this ethics *except* in the quest for *universal* rights?

Thinking along those lines then, forward-thinking humans should probably define the definition of Life so that they may expand the definition of *mercy for Life*, not simply the protection of useful species. Perhaps the reason humans no longer look at trees and flowers with awe is because humans *know them well*—like Hamlet holding the skull of someone he once knew, kissed, and admired? (Shakespeare, 1992). Perhaps by venturing into outer space with this species' leafy companions along as *pals*, humans might renew that lost affection they once held for one another? Humans do seem comforted by any living thing at all, when they find themselves utterly alone—whereas when they are among a throng of conspecifics, they seem uneasy. Researchers argue that humans may have lost their first love for Nature, and that is why they are so uneasy—unloving, unloved—among each other (Costello & Hodson, 2014). Even Antoine St. Exupery, who spent time marooned in the deserts of northern Africa, wrote a children's story about a companion in the shape of a rose—the longing for a single sign of life amid the barren wilds of loneliness (Saint-Exupéry, 1943). Modern studies of plants in space suggest that man's best friend is a begonia rather than a bulldog. Moods are lifted by simply adding a few live plants to a room—why is that, if society goes to such lengths to put every plant and animal safely *outside* its biophobic homes?

In chapter three, the authors delved into the paradox of knowing; the ability to compare one thing to another steals everyone's joy. Why don't humans honor the natural world anywhere near as much as humans honor their knowledge of it? Well, perhaps it's because humans heap things into categories to make them easier to understand—and when they do that, it's really hard to see an endless individuality. The ancient Greeks thought about this too. They called it the Sorites Dilemma. What that was, was a simple set of questions that were very hard to answer. How many grains of salt does it take to make a heap of salt? How many grains can you take away and still have a heap of salt? How many people does it take to make a People? How many roads must a man walk down before he earns the right to be called a man? Must a woman become a man to merit a man's regard? Honor begins when humans allow themselves to grant *unconditional honor* to each individual, not when individuals in a "category" meet the criteria for being in that category. In the same way, reverence for Nature begins when the smallest natural thing is honored—not when Nature must meet conditions to be revered.

Taxonomy is the habit of sticking unique things into heaps and then trying to regard them with the same value outside that heap. The habit of categorizing things steals one's awe for the things one has categorized. Now one is more concerned with where they "go" than what they are. Nature does not need the human mind to classify and record it before it chops Nature down as much as it needs the human heart to declassify and adore it so much its mind would never chop it down. Some of the most famous scientists have stated there is no such thing as a species save by scientific convenience (Darwin, 1876). The reader played a game with this idea, asking how many individuals must die before a species of them is in danger of extinction, and asking how many individuals it takes to make an individual. One of anything humans have never encountered before is a new species, really, so that it isn't very feasible to wait and see how many new species it takes to constitute that one new species. Maybe categories do not exist except as a means for the scientific scouts of humankind to share what they have seen with those back home who haven't? The principle of being both separable (different) and united (similar) is not very sensible. But to fit in nowadays, one must find a way to be both different and the same because everybody has to reconcile this insensibility with common sense (Austen, 1998; Karakostas, 2012). No wonder Nature cannot find fit in human culture . . . Nature must be measurable before humans will admit it exists. Why would the quality called Awe want to be measurable? Awe begins when the slide rule becomes useless.

The first section also spoke of colonialism and expansionism. It discussed how the citizens of a town are more worried about saving their "town" than they are about saving each other. How many townspeople can one lose and still have a town? The answer is *all* of them can be lost. Humans have plenty of ghost towns and buried cities, and all of them still exist even though everybody they symbolized is long gone. Maybe a town *is* its people, not where the buildings are; that certainly seemed true of the diaspora of the Jewish people, who were like a living city that thought they needed a town to be amazing, yet enriched the lives of so many more when they had no common locus at all. So, if taxonomy steals humanity's reverence for things, maybe reverence can restore humanity's awe for things? And maybe, just maybe, when everything goes together well, not certain things, that is when humans will have rediscovered their awe for everything? When they cannot judge what goes with what, or what is more perfect than a similar thing, they suddenly value each individual so highly they cannot compare them to anybody else. If humans can see Nature this way, maybe that will help them see one another with as much reverence?

Chapter four went beyond the taxonomy of species. It went beyond the idea of species in the familiar sense. It proposed that animals are not a different species; it suggested that animals are judged as less than human because humans try to see them as underdeveloped humans instead of fully developed animals. Then that chapter extended this idea and asked whether the children of the human species are judged as less human because adults try to see them as undeveloped grown-ups instead of fully developed children. Maybe childhood is a species, not an inferior stage along the way to becoming an adult? Maybe old age is a species, not a worn-out adult? Maybe both childhood and old age have a worth of their own, with or without ever having achieved fit during middle age? (Meinecke, 2017). This much seems certain: when society waits for something to become something better, it is like telling that thing it doesn't matter in the grand scheme of things until it becomes "better." But when society accepts a thing for what it already is, it is like telling that thing society is satisfied with what it has already become. Human children are being treated as though they are not human enough to merit any reverence, just as women are treated as imperfect men instead of amazing women. But if society could treat both as fully developed species in their own right, then no child would feel inadequate, and no woman would feel inferior. It makes little sense to spend 18 years hoping to be 18, and then spend the next 47 years before the age of 65, hoping to never reach the age of 65. Surely during the years before humans are employed and the years after humans are employed, they are every bit as human as the years in between? Therefore, every human of every age deserves the very same regard from all humans, no matter what epoch of the lifespan he or she might happen to occupy. As Rogers (1961) phrased it, the goal of psychotherapy is to create an environment of understanding, acceptance, warmth, and sincerity. When therapists do that, the client does the same within himself (or herself). If human society did this with its planet, the planet might find its worth again. Both natural children and this natural environment deserve *unconditional positive regard*, rather than demanding they both grow up before society will regard them well.

In chapter four, the authors proposed that *every moment matters*. Just as every species matters (chapter 3), and just as every stage of the lifespan matters (this chapter), every moment everywhere matters. Every stage of Life is the best stage; being one's best is better than becoming one's best. Why do children weep when they should be laughing? Why does grief hurt so much in one's final years? Humans spend their childhood wishing they were older and their golden years wishing they were young again. In the middle, which should be the ideal age by these arguments, humans wish they were either younger or older than this eternal age of employability. If humans honored childhood and old age, perhaps they would honor Nature too? Going the other way, if humans honored Nature no matter how ripe or infertile, perhaps those marginalized individuals that society regards as unripened or infertile would find *an unusual honor* in being what they already are, not what society waits for them to become (Chesnutt, 1968).

RECAP OF SECTION II: PROTECTING THE ENVIRONMENT

In the second section, the examination of biophilia indicates that it isn't just an attitude. The change in attitude can bring about the desire to change the world. There is often the possibility of a healing connection between the natural world and the human world, when the human world remembers to reach back and become acquainted with its ancient relationship (and ongoing indigenous relationship) with the natural world. In chapter 5, it became clear that biophilia can help ailing human populations feel grounded, centered, focused and reassured that everything will be okay. Just beyond the concrete forests humans usually navigate as they go to work and back, still stand the cities of God in the form of forest skyscrapers and animal traffic below. Humans can apply biophilia any day at all by venturing out into a group session with lakes and trees and birds and clouds, the communication with which humans can find therapeutic time not available among hermit crab-like vehicles and the wave-like roar of industry. One kind of therapy, Shinrin-yoku (forest therapy) has already shown promise. In fact, the very idea of living in densely populated areas can bring about distressing biopsychosocial effects reminiscent of withdrawing from Nature (maternal care), or trying to get by like a runaway child (self-care).

And instead of leaving town for therapy, humans can also issue visas to Nature and let it dwell contentedly among humans in town, side by side—with green spaces and the benefit of positive natural stimuli. Exposure to visual and auditory pollution can be partially offset by interrupting the regularity with some natural foliage and natural noises. City-dwellers can *plant with purpose*, which means not so much for agriculture but to coexist with types of living things that restore one's sense of belonging to the world, and caring for Nature for its own sake. Food that is typically discarded as waste may be re-envisioned as potential gleanings, just like ancient civilizations who shared their bounty with the poor. A change in attitude toward plants may also involve a change in attitude toward the animals and insect species that collaborate with plants, such as bees. Or, rather than planting token trees or privacy hedges, people may plant with purpose by creating labyrinths for recreational relief. In such cases, sometimes people discover a sacred geometry hidden in their natural relationship between greenery and people, adding more than just color to the healing impact of foliage—a sort of lunar effect right down here on Earth, whenever people see cyclic or tidal nuances in natural phenomena.

In chapter six, the authors proposed a new theory. They called it *National Park Theory*. The theory is simply that, if humans could just treat their world as visitors again, not owners, they might rediscover their reverence for Life. Visitors to Mankind's National Parks come to appreciate the wonder and sheer majesty of Nature—not to figure out ways to profit from those wonders and leave tree stumps and mines behind. People from all over the world come to Zion, Utah, for instance, just to be near these awesome sights for a week or so. (One of the authors is lucky enough to live here!) Beneath the idea of awe is another word for appreciation, termed *regard*. There are first impressions that stay with people whether they want to remember or not, and lasting impressions that make them long to go back. Neuroscientists teach that there probably are no actual patterns in Nature; rather, a pattern helps humans return to whatever experience amazed them, or stay clear of whatever experience terrified them. A pattern is like a *just noticeable difference* that is difficult to avoid seeing, once someone has seen it.

Human attitudes toward the planet and one another are often very *feedforward*, meaning they make up their minds without ever looking again to see if their opinions are biased or flawed. Why would humans do this? Perhaps it's because their thoughts are jealous of the things they refer to? There is some evidence—Nobel Prize winning evidence—that neurons "care" whether they are perceived as useful by the brain they occupy, for if they cease to be useful for their typical function, they must allow themselves to be recruited for another function or be deprived of blood and oxygen, and inevitably die (Wiesel, 1982). A bias is still made of living cells, and if the bias is no longer needed, neither are the cells. Perhaps the odd fury of being disagreed with stems from this physiological explanation, not stubbornness or ignorance. In other words, everything either matters or it will become extinct.

A key concept, in addition to National Park Theory, was the possibility that the human mind views its own body the same way it views its planet. If it views its body as a means to achieve status and pass down its wealth, it will probably view the planet the same way.

In chapter seven, this book took a slight detour from the positive aspects of reverence for Nature, to the exploitation of reverence for Nature during times of economic distress. Humans may mean well, but often the vulnerable become a means for the powerful to take advantage of their sympathy. Environmental protection sounds like a good idea, but recently it seems to have become more about protecting commercial interests than about protecting the environment's survival. If a thing is beloved, the need to protect it should not come to mind (unless most humans doubt that others will love it as much as a few humans do). That doubt is a projection of humankind's own failure to love the beloved object completely, and have faith that its beloved can make any heart surrender without a plea. This same phenomenon happens when a man falls in love with a woman, but later forgets how easy it was to fall in love—affection is its own shield, and it inhibits harm because anyone would be enchanted by it, not because some folks own it and guard it well. Such affection is affection no matter what; such faith is faith no matter what; such love is love no matter what. Beyond affection or environmental protection, the notion of beneficence has been misused too, since the word means kindness but its definition is usually mistaken for maximizing one's own benefit while minimizing the object's harm. That definition is not about kindness, but how to exploit a thing without looking bad.

This section provided an example of the widespread impersonation of sympathy, with the ostensible appeal to save the wildlife and beauty of an Alaskan bay while really trying to protect the fishing industry there (which was already exploiting the wildlife). It was learned that this habit of putting human commercial interests first and Nature's well-being second has been around a very long time—at the dawn of the Middle Ages. It also shared some examples of a *scorn for juvenility* by humankind, in that humans look at children of any species as an unprofitable epoch of the lifespan (unless they mature into profitable enterprises). So if this kind can learn to see every other kind at any age of its development with reverence, perhaps this species might take a second look at its own children, and reassure them that they do not have to grow up to be rich and famous to be the starring roles in their parents' lives lifelong. If anything, human compassion and love for each other increase when one needs constant attention and the other realizes he or she has the privilege of being the other's caregiver. That translates well to juvenility, because it can feel like an unrewarding career to raise a child who turns out badly—but it is always rewarding if the goal of raising a child is simply to love them no matter what. The fact that parents or friends had this time together to be of some help to someone they adore, should be sufficient reward—yet if people keep track of how much or how well they served, for some reason they forget how grateful they already were. Speaking of mistreatment, this chapter recalled how the plague of colonialism once viewed unconquered lands as child-like and backward, and used the idea of immaturity as an excuse to rob and occupy their territory in order to improve their lot.

The middle theme of the book concluded with a look at dominion. Societal indifference to environmental protection is matched only by the view that some things need to be dominated, which is an indifference to their own indefeasible right to remain free and be treated as equals. Humans have always assumed they should decide for those who cannot make their own decisions—even if what authorized humans decide on their behalf is plainly harmful and unfair, and not something the represented group would ever decide to do. This is true for juveniles, the elderly, the physically disabled, and the mentally infirm—and for Nature, whether it is juvenile, elderly, physically ailing, or perfectly healthy (because it has no mental aspect at all). Human civilization uses unnatural reasoning and unnatural methods to control natural populations while doing little about its own; it allows animal populations to be decimated for sport when humans would never allow this behavior to control their own population, no matter how dire the circumstances. Such differential logic suggests insincere motives and selective application of justifications in the peculiar interests of just one dominant species. Across time, the dominion model, royal model, colonial model, and managerial model have been used to justify the involuntary reduction of otherwise choice-capable species who would not vote for such practices, were they given the right to assembly and peer representation. Mostly, humans have not changed their warden/warren mindset, but only broadened the membership of elite hunters and worsened the predicament of the hapless animals whose lives and families are transformed into seasonal nightmares—who cannot trust the fickle motives of their cohabitants and protectors.

Ecosystem management is another misused phrase, in which human meddling with naturally regulated living systems is always disastrous. The intelligent animal does not seem to be able to think as most animals do, and because humans cannot think like they do, everything humans "think" is best for them, is eventually tragic to these non-mentally directed forms of life. What is best for the mental species is plainly the death knell for every other species, with an explosion of habitat destruction, planetary pollution, and (instead of considering how to save the planet), solutions that include abandoning the dying Earth to gleefully devastate another world (Wilson, 2016). Perhaps what is needed is to observe how animals survive human destruction of the environment, and learn from them? Perhaps the answer is not more effective management, but to cede sovereignty back to the natural monarchy which has survived without a human sovereign far longer than Mankind.

RECAP OF SECTION III: ADDRESSING SOCIETAL INDIFFERENCE

The theme of the last four chapters (this section) has centered on practical ways to help restore the Earth to its former glory—things everyone can do right now. This book did not focus on the all-too-common solicitation of political contributions from casual consumers, in the hope the public can passively influence lawmakers to include the planet it lives on in the idea of representation. Saving the planet is not about who gets to save it. If this proposal succeeds, it will not matter who saved it or how, since no one will be left to judge such a miracle.

The first of these last four chapters introduced the *Hurricane Survivor Effect*. In the authors' research, they discovered that when people lose a little of their many possessions due to a disaster, they become unexpectedly upset as though they were cheated out of their share (even if they are otherwise nice people). They grumble over having to make repairs instead of being grateful to have survived. But when they lose everything to a disaster—even though one would expect them to be even more upset—instead they become surprisingly humble, and grateful to be alive. They suddenly appreciate their families and friends, now that all of their material goods have been lost. It does not seem to be a character quality, so much as the serendipitous ability to finally see what really mattered to them. So it was proposed that, if humans could apply this principle to the loss of the entire planet (before it happens), they might treat it with the same gratitude as their own families, when they lose everything except them. So what if humans lose their jobs? So what if humans lose everything its race has accomplished? How much luckier could humans be, than to realize (when Mankind has finally lost the last of its mighty changes), that this species has saved its beloved world in that endeavor? Climate change is a sort of hurricane survivor effect… a wakeup call to cherish the natural world (which is about to die), not cherish human industries (which are the reason the world is about to die). Even if this species should lose its homes and its cities, it may come to realize what was more important all along—a place to put them.

In the second of these last four chapters, the authors proposed that modern civilization should view this dire situation (between the Earth and Mankind) as an opportunity for a new kind of conflict resolution (a skill at which one of the authors is an expert). The Biophilic Model of Conflict Resolution would recognize Nature's rights; it would recognize the rights *of* Nature (not Man's rights *to* Nature). Humanity's connection to Nature extends far beyond legal justice and responsibility (Carli, 2018). Current practices (in resolving conflicts between this kind and more natural kinds) are predicated on outdated belief systems, such as mechanistic, anthropocentric, and adversarial paradigms. Nature needs to be awarded the status of *personhood*—not simply elevated to the status of priceless real estate. It has been a respect for all persons no matter their financial status, which freed humans from figuring out whether they could do without a human individual financially, and still maintain the status quo. But it will take more than clever legislation to represent the rights of a natural system that dwarfs the nature of policy. Mother Nature does not negotiate contracts—it simply wants to survive, like any living soul. Contracts do not have souls, so far as anyone can tell.

In the third of these last four chapters, this book about biophilia introduced a novel yet practical way to increase human reverence for Life. If humans could only see a single wild rose as the most precious thing on Earth, promise to do absolutely anything to save its rose, imagine how they might see each other? Imagine how they might value the planet? Can reverence for a flower really save a dying planet? Yes it can. Yes it can. Repeat after Life: yes it can. Reverence for Life is a very powerful feeling. And when a person treasures the *slightest glimpse* of Life, that person becomes *hypervigilant* to Life everywhere he or she goes. If great men would only kneel to a flower as though to a coming king, that kingdom might come—a natural kingdom in which all living things were blessed, not just things shaped like people.

This seems like a good place to end a book. The authors hope they have inspired you to pause along the way, and whisper "I love you!" to the first living thing you see. It may not be human; it may not understand your language. But it will surely understand that you stopped doing everything else you might have done, to let it know how much it mattered to you. If human society could only do the same for all living things, someone or something would surely notice how much that brief society cared.

REFERENCES

Austen, J. (1998). *Sense and sensibility*. London, UK: Wordsworth Editions Ltd.

Bradbury, R. (2001). *The illustrated man*. New York, NY: Book-Of-The-Month Club/Avon Books.

Carli, J. (2018). *Legal rights for nature: A revolution whose time has come*. Retrieved from https://medium.com/@jamescarli/legal-rights-for-nature-a-revolution-whose-time-has-come-fd11c65545cf

Chesnutt, C. W. (1968). *The house behind the cedars*. Ridgewood, NJ: Gregg Press.

Cohrs, J. C., Petzel, T., & Funke, F. (2012). Authoritarian personality. In D. J. Christie (Ed.), *The Encyclopedia of Peace Psychology* (Vol. 1, pp. 62–66). doi:10.1002/9780470672532.wbepp016

Costello, K., & Hodson, G. (2014). Explaining dehumanization among children: The interspecies model of prejudice. *British Journal of Social Psychology, 53*(1), 175–197. doi:10.1111/bjso.12016 PMID:23083220

Darling, N. (1999). *Parenting style and its correlates*. ERIC Digest. Retrieved from ERIC database. (ED427896)

Darwin, C. R. (1876). *The origin of species by means of natural selection, or the preservation of favoured races in the struggle for life* (6th ed.). London, UK: John Murray.

Deleuze, G., & Guattari, F. (1987). *A thousand plateaus: Capitalism and schizophrenia*. London: Continuum.

Freud, S. (1957). On narcissism: An introduction. In J. Strachey (Ed.), *The standard edition of the complete psychological works of Sigmund Freud* (Vol. 14, pp. 73–102). London: Hogarth Press.

Gilhus, I. S. (2006). *Animals, gods, and humans: Changing attitudes to animals in Greek, Roman, and early Christian thought*. London, UK: Routledge.

Goldberg, M. (2009). *The means of reproduction: Sex, power, and the future of the world*. New York: Penguin Press.

Gopnik, A. (2016). *The Gardener and the carpenter: What the new science of child development tells us about the relationship between parents and children*. New York, NY: Farrar, Straus, and Giroux.

Gould, K. (2011). First love [Recorded by K. Gould]. On Roots of Love [CD]. Nashville, TN: Fine Linen Music.

Grandin, T., & Johnson, C. (2005). *Animals in translation: Using the mysteries of autism to decode animal behavior*. New York, NY: Houghton Mifflin Harcourt.

Hansberry, L. (1959). *A raisin in the sun: A drama in three acts*. New York: Random House.

Irvine, A. D. (2019). Bertrand Russell. In E. N. Zalta (Ed.), *The Stanford Encyclopedia of Philosophy*. Retrieved from https://plato.stanford.edu/archives/sum2019/entries/russell/

Jetten, J., & Mols, F. (2014). 50:50 hindsight: Appreciating anew the contributions of Milgram's obedience experiments. *The Journal of Social Issues*, *70*(3), 587–602. doi:10.1111/josi.12080

Karakostas, V. (2012). Realism and objectivism in quantum mechanics. *Journal for General Philosophy of Science*, *43*(1), 1–20. doi:10.100710838-012-9173-5

Keltner, D. (2009). *Born to be good: The science of a meaningful life*. New York, NY: W.W. Norton & Co.

Meinecke, L. D. (2017). *Neglected by assessment: Industry versus inferiority in the competition for scarce kidneys* (Doctoral dissertation). Available from ProQuest Dissertations and Theses database. (ProQuest No. 10689852)

Meinecke, L. D. (2018a). A theory of cognitive idolatry, or the human struggle to preserve psychological certainty in a biological species - chapter 1 (Research proposal). doi:10.13140/RG.2.2.16527.43684

Meinecke, L. D. (2018b). A theory of cognitive idolatry, or the human struggle to preserve psychological certainty in a biological species - chapter 2 (Research proposal). doi:10.13140/RG.2.2.31089.53601

Meinecke, L. D. (2018c). *Hide her under a bushel? No, I'm gonna let her shine...* Retrieved on March 12. 2020 from https://www.linkedin.com/pulse/hide-her-under-bushel-im-gonna-let-shine-lonny-meinecke-phd/

Mitchell, M., & Conroy, P. (2011). *Gone with the wind.* New York: Scribner.

NASA. (n.d.). *JFK Rice moon speech.* Retrieved from https://er.jsc.nasa.gov/seh/ricetalk.htm

Newlands, S. (2014). Leibniz on privations, limitations, and the metaphysics of evil. *Journal of the History of Philosophy, 52*(2), 281–308. doi:10.1353/hph.2014.0033

Newman, J. (2014). Hannah Arendt: Radical evil, radical hope. *European Judaism, 47*(1), 60–71. doi:10.3167/ej.2014.47.01.07

O'Grady, M. A. (2016). *The theory of biophilia* (Reg. No. TXu 2-032-922). United States Copyright Office. Retrieved from https://www.copyright.gov/

O'Grady, M. A. (n.d.). *The theory of astrophilia* [Copyright pending]. United States Copyright Office. Retrieved from https://www.copyright.gov/

O'Grady, M. A., & Meinecke, L. (2015). Silence: Because what's missing is too absent to ignore. *Journal of Societal and Cultural Research, 1*(1), 1–25. Retrieved from https://www.researchgate.net/publication/271842302_SILENCE_BECAUSE_WHAT'S_MISSING_IS_TOO_ABSENT_TO_IGNORE

Rieber, R. (1983). The psychology of language and thought: Noam Chomsky interviewed by Robert W. Rieber. In R. W. Rieber (Ed.), *Dialogues on the Psychology of Language and Thought.* New York, NY: Plenum Press. doi:10.1007/978-1-4615-9308-9_2

Rogers, C. R. (1961). *On becoming a person.* Boston, MA: Houghton Mifflin.

Saint-Exupéry, A. (1943). *The Little Prince.* New York, NY: Harcourt, Brace & World.

Shakespeare, W. (1992). The tragedy of Hamlet, prince of Denmark (New Folger's ed.). New York: Washington Square Press/Pocket Books.

Steinbeck, J. (2003). *East of Eden.* New York: Viking.

United Nations. (1948). *Universal declaration of human rights.* Retrieved from https://www.un.org/

Wanderer, E. M. (2015). Biologies of betrayal: Judas goats and sacrificial mice on the margins of Mexico. *Biosocieties, 10*(1), 1–23. doi:10.1057/biosoc.2014.13

Wiesel, T. N. (1982). The postnatal development of the visual cortex and the influence of environment. *Bioscience Reports*, *2*(6), 351–377. doi:10.1007/BF01119299 PMID:7049262

Wilson, E. O. (2016). *Half-earth: Our planet's fight for life*. New York, NY: Liveright Publishing.

Wise, R. (Producer/Director). (1965). The sound of music. Beverly Hills, CA: Twentieth Century Fox Home Entertainment.

Conclusion:
Awed by Nature

The authors have covered a great deal of territory over the course of these 12 book chapters, but in reality, that progress was made by transitioning from the theoretical to the practical application of biophilia. Initially, the objective of the first section of this book was to introduce the audience to *The Theory of Biophilia* so that readers might gain a better understanding of just how close their relationship with Nature actually is. From that introduction, the authors guided the reader toward a deeper understand of biophilia—more like a point of total immersion (internally and externally), wherein Nature exists within and without every individual, human and non-human, on this planet.

Perhaps even a bit more thought-provoking was the introduction of *The Theory of Astrophilia*™ which expanded the premise of biophilia beyond Earth's gravity, because to believe that humans are alone in the Universe would seem to assume a very egocentric perspective indeed. However, the human proclivity for egocentrism was also presented in chapters three and four as these authors discussed how humans have constructed taxonomy as a means of categorizing humans and non-humans for the sake of convenience, while at the same time stealing their uniqueness by lumping them into groups, heaps, and the like. The further lack of human reverence for living things was evidenced by the content of chapter four, in which the value of childhood and the wisdom of old age was shown to be sadly lacking. Children are chided for not growing up quickly enough to emulate adults while they are still children; and seniors are criticized for draining resources needed by the mentally and physically fit—no longer productive members of a hard-working society, and too old to grow up and become productive..

The second section of this book discussed the pros and cons of protecting the environment on Earth as well as the underlying human motives for that so-called "protection." Instead of simply recognizing how nurturing Nature can be when humans treat Her with love and respect, humans are vigilant to how to benefit from her mistreatment. Many cultures have recognized the healing and grounding potential of Biophilia, as demonstrated by Shinrin-Yoku which can be translated as "forest bathing" as it is practiced in Japan. Many indigenous peoples have long known about the healing capacity of Nature not only in her medicinal herbs and plants, but also in Her capacity to heal and refresh the mind and spirit in addition to the body.

Humans may set aside areas and tracts of land as preserves, national forests, etc. which are supposed to protect them from further encroachment, decimation, and other undesirable human activities; however, the entities making such decisions as well as the decisions regarding who is guarding them may just be protecting them into extinction. Who but Nature herself is wise enough to self-heal if left to Her own devices, and without further human interference? Additionally, humans are looking ever skyward with the potential capability of inter-stellar space travel, as well as the search for exoplanets for colonization. This suggests that when the Earth no longer has sufficient resources to share or harvest, humans will be "jumping ship" to colonize other planets and to begin the entire process all over again on another unsuspecting world.

The final section assumed a most unique perspective as it relates to humans' inextricable link to Nature and its non-human citizens. A "what if" scenario was presented, like a tipping point of being able to save the Earth before it is too late, as the inexorable demise of Life itself looms ever closer. What if humans altered their perspective from how much (or perhaps better said, how little) they "must" give back to Nature in an attempt to save the planet, to a perspective of cheerfully volunteering to give everything back to Nature?

What if humans regarded their relationship with Nature as a collaboration as opposed to a competition for resources? What if humankind had no burning desire to dominate everything else, because that desire was about love and not about domination? What if humans began to comprehend the concept of human nature in its most basic form—which is the concept of expressing a reverence and a love for everything in Nature down to the simplest flower? What if humans could transcend their perspective of self-love, and imagine that there is no greater expression of love than the willingness to give up everything for the survival of everything?

About the Authors

Mary Ann Markey is an Adjunct Professor at Grand Canyon University with a PhD in Conflict and Crisis Management and a Master's Degree in Psychology. Over the course of her professional career, her research, presentation and publishing topics have included: biophilia, astrophilia, intra-family homicide, revenge, violence, aggression, domestic violence, torture, active killers in schools and workplaces, and mass shootings at public places and social venues (movie theaters, music concerts, etc.). Website- https://www.gemconferences.com

Lonny Douglas Meinecke, is an Adjunct Instructor at King University with a PhD in Psychology. He researches marginal topics concerning social justice (ageism, ableism, and animal justice), and is a member of several organizations including the Social Psychology Network, the Psychology Coalition at the United Nations, the Task Force on Indigenous Psychology, the Society for Theoretical & Philosophical Psychology, and the International Society for Justice Research. His work is about promoting reverence for childhood and old age, and the importance of honoring biological life of every age and every kind, while Life is briefly here. Website- https://lmeinecke.com/2020/

Index

I

L

M

N

O

P

R

S

T

W

www.ingramcontent.com/pod-product-compliance
Lightning Source LLC
LaVergne TN
LVHW081329300926
849161LV00008B/1920
* 9 7 8 1 7 9 9 8 4 4 0 8 2 *